Optimization of Hydraulic Fracture Stages and Sequencing in Unconventional Formations

Optimization of Hydraulic Fracture Stages and Sequencing in Unconventional Formations

Dr. Ahmed Alzahabi
Dr. Mohamed Y. Soliman

CRC Press
Taylor & Francis Group
Boca Raton London New York

CRC Press is an imprint of the
Taylor & Francis Group, an **Informa** business

CRC Press
Taylor & Francis Group
6000 Broken Sound Parkway NW, Suite 300
Boca Raton, FL 33487-2742

First issued in paperback 2021

ISBN-13: 978-0-367-78106-4 (pbk)
ISBN-13: 978-1-138-08595-4 (hbk)

Library of Congress Cataloging-in-Publication Data

Names: Alzahabi, Ahmed, author. | Soliman, Mohamed Y., author.
Title: Optimization of hydraulic fracture stages and sequencing in
unconventional formations / Dr. Ahmed Alzahabi, Dr. Mohamed Y. Soliman.
Description: Boca Raton, FL : Taylor & Francis Group, LLC, [2018] | "CRC Press
is an imprint of Taylor & Francis Group, an Informa business." | Includes
bibliographical references and index.
Identifiers: LCCN 2018009766| ISBN 9781138085954 (hardback : acid-free paper)
| ISBN 9781315111162 (ebook)
Subjects: LCSH: Gas wells--Hydraulic fracturing. | Oil wells--Hydraulic
fracturing. | Petroleum--Geology.
Classification: LCC TN871.255 .A49 2018 | DDC 622/.3381--dc23
LC record available at https://lccn.loc.gov/2018009766

Visit the Taylor & Francis Web site at
http://www.taylorandfrancis.com

and the CRC Press Web site at
http://www.crcpress.com

Contents

Acknowledgments

The authors would like to acknowledge with great appreciation the theoretical and practical ideas from the following evaluators and contributors that, through complicated models, enriched this work.

Professor George Asquith
Texas Tech University

Professor Richard Bateman
Texas Tech University

Professor Alex Trindade
Texas Tech University

Mr. Tim Beims
American Oil and Gas Reporter

Dr. Tim Spinner
Marathon Petroleum

Dr. Noah Berlow
Texas Tech University

Dr. Ghazi Al-Qahtani
Saudi Aramco

Professor Ismael De-Farias
Texas Tech University

Mr. Neil A. Stegent
Pinnacle Technologies Inc.

Dr. Ravi Vadapalli
Texas Tech University

Professor Ali Daneshy
Daneshy Consultants

Mr. Buddy McDaniel
Halliburton

Their comments have greatly helped in the optimization aspect of the book and led to improvement of the quality of the book. The authors thank Buddy McDaniel from Halliburton for authoring Chapter 1 of this book "Fracturing Chronology: Milestones of the Hydraulic Fracturing Process" in Chapter 1 of this book. Finally, we would like to express our appreciation to Neil A. Stegent, from Pinnacle Technologies Inc. For advising in the completion aspect of Chapter 1

Ahmed Alzahabi
Mohamed Y. Soliman

Authors

Dr. Ahmed Alzahabi is an assistant professor at University of Texas of the Permian Basin. He earned a PhD and a MS engineering from Texas Tech University, an MS from Cairo University and BS from Al-Azhar University; all in petroleum engineering. He previously served as a researcher at the Energy Industry Partnerships of University of Houston. His research involves developing techniques for Permian Wolfcamp exploitation. He has participated in six US patent applications and edited and reviewed for multiple journals. He has contributed a book chapter in Fracturing Horizontal wells and a book on PVT Property Correlations.

Dr. Mohamed Y. Soliman is department chair and the William C. Miller endowed chair professor of petroleum engineering at the University of Houston. He received his PhD from Stanford University in 1979. He is a distinguished member of SPE and a licensed professional engineer by the State of Texas. He is also a fellow of the National Academy of Inventors (NAI). He has authored and coauthored more 200 technical papers and holds 29 US patents. He is also the editor of *Fracturing Horizontal Wells*, published by McGraw-Hill in July 2016. His areas of interest include well test analysis, diagnostic testing, fracturing, and numerical simulation.

1

Fracturing Chronology: Milestones of the Hydraulic Fracturing Process

1.1 Motivation and Objective

More than 70 organic-rich shale oil and gas plays have been identified to date in North America (Wang and Gale 2009). Many of these extremely low-permeability "source rock" formations have wet gas and/or condensates associated with the production streams, with parts of some plays often yielding more dry gas and other parts more prone to natural gas liquids and/or oil.

Shale gas and oil play identification are subject to many screening processes for characteristics such as porosity, permeability, and brittleness. In fact, evaluating shale gas and oil reservoirs and identifying potential sweet spots requires taking into consideration multiple rock, reservoir, and geological parameters that govern production.

The early determination of sweet spots (portions of the reservoir rock that have high-quality kerogen content and brittle rock) for well site selection and fracturing in shale reservoirs is a challenge for many operators. Often, only certain parameters are used (i.e., only brittleness) to determine the sweet spots for well site selection. Additionally, the fractures are generally placed equidistantly. This may lead to short transverse and/or axial fractures that are problematic during hydraulic fracturing and may not result in optimal production.

With this limitation in mind, an approach has been developed as part of this book to improve the industry's ability to evaluate shale gas and oil plays. The approach uses a new candidate selection and evaluation algorithm and screening criteria based on a number of key geomechanical, petrophysical, and geochemical parameters and indices to obtain results consistent with existing shale plays and gain insights on the best development strategies going forward.

Specific parameters include thickness, depth, total organic carbon content, thermal maturity, brittleness, mineral composition, total porosity, net thickness, adsorbed gas, gas content, and geologic age. A database of these properties from 12 major North American shale plays (Barnett, Ohio, Antrim, New Albany,

Lewis, Fayetteville, Haynesville, Eagle Ford, Marcellus, Woodford, Bakken, and Horn River) was established to guide the algorithm considering all the properties and potential approaches to future reservoir development.

The efficient exploitation and development of shale plays comes from application of the appropriate technology (e.g., horizontal wells and multistage fracturing, both in the right place and at the right time) to reach the sweet spots of unconventional reservoirs. This result is possible through properly integrating both geochemical and geomechanical analysis. Employing such technology facilitates the selection of the best methods for future shale play development while minimizing bad wells (wells that cannot produce at economical rates), implementing better completion strategies, enabling a more effective fracturing campaign, and resulting in optimized production and reservoir management.

Recently, it has become clear in the industry that multistage fracturing is the approach most often used to drain shale reservoirs (Soliman et al. 1999). The problem is that not all stages contribute to production, according to Miller et al. (2011). Miller analyzed 100 horizontal shale wells in multiple basins and found that two-thirds of total production comes from only one-third of the perforation clusters.

Use of three-dimensional (3D) seismic and sonic data is the best method to represent the complex shale model. Miller et al. (2011) recommended employing reservoir and completion quality in designing stages and clusters along horizontal wells. Reservoir quality is defined by petrophysical properties of organic shale that make it variable for development, such as maturity, porosity, and organic content. Completion quality is defined by geomechanical parameters that are required to effectively stimulate the shale; these parameters include stresses, mineralogy, and orientation of natural fractures.

Currently, the shale brittleness indicator (BI) is used to identify more brittle and more productive shale (Jarvie et al. 2007; Rickman et al. 2008; Jin et al. 2014). Rickman et al. (2008) defined brittleness as a function of Young's modulus and Poisson's ratio, attributing the success of any fracturing placement to geochemical analysis; the analysis can be obtained through petrophysics and lab measurements. Their work was more closely related to shale; this study considers productive segments of horizontal wells versus unproductive ones with exact limits of the cutoffs between ductile and brittle rock (the start and end of each rock classification on an approved scale such as fracturability index [FI]).

Well site selection and fracture placement in shale are not easy decisions. The decision of whether to initiate a fracture stage must be made frequently. Mathematical programming decision variables are used to represent yes/no decisions to fracture. At the same time, maximizing the values of the selected good positions for wells or fracturing is trivial, especially in light of the need to respect certain constraints on possible deviated well locations and fracturing positions. These constraints of the given problem include minimum/maximum well spacing and distance between the fracturing locations in the same deviated well. Important factors to be addressed include stress reversal criteria and the ordering of places with the highest fracturability without consideration of fracture creation time.

This work will introduce new criteria that will accurately guide the development process in unconventional reservoirs in addition to reducing uncertainty and cost. Previous studies that yielded successful results for conventional wells will not be applicable to unconventional resources, as shale has unique characteristics such as ultrapermeability and the presence of extensive natural fractures; therefore, application of potential productivity maps (PPMs) or so called geo-bodies is not effective. Due to the severe heterogeneous environment in shale gas, methods such as reservoir quality (RQ) may not be valid for driving the selection of sweet-spot points. (Reservoir quality was introduced by Vasantharajan and Cullickin (1997).) Stegent et al. (2012) used an engineering approach combining elevated factor vanadium (EFV) as an indicator of total organic carbon and relative brittleness index (RBI) for selecting intervals that make initiating fractures easy. They applied their technique for optimizing shale oil from Eagle Ford.

1.2 Book Outline

The book is structured to lead the reader from general shale oil and gas characteristics to detailed sweet-spot classifications. It is organized as follows.

Chapter 2 provides a comprehensive literature review with a focus on a new approach developed for screening criteria for shale gas and oil plays. It highlights how a total of 15 parameters consisting of geochemical, petrophysical, and geomechanical mapping are the best tools for screening the potential of shale plays, and addresses the challenges associated with the identified methods. It proves that statistical similarity and clustering analysis techniques may reveal previously unknown relationships among the 12 shales. It also targets a comprehensive predictive model to evaluate the key success of completion strategies (treatment) for the major successful shale plays and guide future selective optimum completion for each shale play.

Chapter 3 shows a fracturability index as a recently developed concept for identifying sweet spots in shale reservoirs. It is based on geomechanical principles and has been developed to optimize the placement of fractures along horizontal and deviated wells in unconventional reservoirs. Based on the index, an algorithm has been developed to prioritize the brittle and high in-situ stress zones along the well path. The algorithm suggests the order of possible fracture locations for future resource development in single or multiple wells. The order reflects a ranking of fracture stage placement according to their potential and not fracture creation time.

Chapter 4 introduces another new sweet-spot identification method known as mineralogical index maps. The developed method, which considers mineralogical parameters, is a new sweet-spot identifier that guides well placement and hydraulic fracturing positioning in unconventional resources. Well performance in a form of a total organic content total organic carbon (TOC) comparison is conducted between the proposed method and previously published techniques, such as total organic content, by combining the mathematical optimization with each method. This chapter was published in the form of a paper in SPE-178033-MS.

Chapter 5 introduces the design and validation of a new integrated combined fracturability index correlation. The new integrated FI takes both geomechanical and geochemical effects into consideration. The new classification identifies the shale reservoir based on brittleness, high porosity, and organic material.

Chapter 6 considers several computational techniques for solving one formulation of the well placement problem (WPP). Typically, the well placement problem is approached through the combined efforts of many teams using conventional methods, which include gathering seismic and sonic data, conducting real-time surveys, and performing production interpretations to define the sweet spots. This work considers one formulation of the well placement problem in heterogeneous reservoirs with constraints on interwell spacing. The performance of three different types of algorithms for optimizing the well placement problem is compared: genetic algorithm, simulated annealing, and mixed integer programming (IP). Example case studies show that integer programming is the most accurate approach in terms of reaching the global optimum solution.

Chapter 7 discusses the use of IP in placing horizontal wells in the most productive segments of the reservoir. We place fractures in an overlapping or staggered design to reduce the stress-shadowing effect and thus obtain optimal fracture geometry and improve overall expected reservoir production.

Chapter 8 introduces multigrid fracture-stimulated reservoir volume mapping coupled with a mathematical optimization approach to shale reservoir well and fracture design, scheduling, and development.

Chapter 9 presents the conclusions of this work and recommendations for future work.

1.3 Book Chronology: Milestones of the Hydraulic Fracturing Process

In 2006, US oil production had fallen to modern-day record lows, and domestic consumption depended heavily on imported oil from Organization of the Petroleum Exporting Countries (OPEC) countries as well as Mexico and Canada. By late 2007, prices for crude eventually closed in on USD 110 per barrel (2013 dollars); however, in 2008, the US "real estate bubble" burst, and excessive subprime mortgage lending brought many banking failures, choking our economic growth and fast-forwarding us to a global recession.

However, just before this, something quite unexpected was already beginning to happen within the US oilfield: we proved that some of our well-known resource shale formations still had meaningful quantities of gas or oil in place but needed new methods to make them commercial drilling targets. Horizontal drilling and a new approach to an old innovation called hydraulic fracturing (i.e., with multistage treatment designs placing millions of pounds of sand per lateral drilled) were used in very specific ways. With this innovative approach, the US domestic production of oil and gas rebounded, and by 2017 it had doubled since 2006. Because of the resilience of the oilfield, the US reclaimed the throne of being the world's largest oil producer in 2016, surpassing both Russia and Saudi Arabia. Looking forward, only future economic recessions would seem to be any reason to believe that the potential for future gains would not continue while new fields are discovered and methods continue to improve.

Chapter 1 discusses the 70-year chronology of the hydraulic fracturing method of oil and gas well stimulation.

1.3.1 Introduction

Following the initial idea developments through field trials developing the hydraulic fracturing process between 1947 and 1948, the industry followed with a decade of extensive commercial application, but with few significant improvements on the process in this early time period. During the six decades after this initial development era, we have witnessed wide acceptance and proliferation of hydraulic fracturing as a functional well stimulation process, with many scientific milestone developments from within the fracturing technology community. There have even been several truly scientific projects that were usually able to help the industry better comprehend what had occurred and better predict what can happen downhole during the pumping of a fracturing treatment. However, as with all industries, the fracturing stimulation community also benefited greatly by the worldwide growth of communication, technologies, and technical tools.

From a big-picture point of view, drilling activity levels (and therefore the volume of fracturing stimulation work) can generally be described as "chasing the oil and gas price." Of course, there is some lag to starting and ending most of these price chases since enlarging equipment fleets and hiring/training manpower are far less instant than the occurrence of price fluctuations. Layoffs occur much faster after the oil price declines, and many of those impacted may leave the oilfield forever, which prolongs adequate rehiring and enhances the need for training new hires when activity increases. Some of these ups and downs in activity levels will be discussed, such as when the increased activity focus may be only on oil or gas. We see that often the hills and valleys of the price curves are misaligned, with our industry chasing what is on the rise as

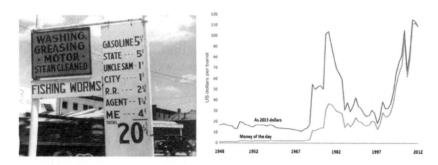

FIGURE 1.1
(Left) US gasoline cost breakdown at a gas station in the late 1940s. (From Google, free to use.) (Right) Yearly average US oil price data from 1948 through 2013 presented in both "dollars of the day" and as 2013 dollars. (From data downloaded from energy information administration (EIA) public website.)

we move our activity away from the falling one to the one where the money now looks better. Woe be unto us when both oil and gas prices are low!

Figure 1.1 provides a representation of the typical late-1940s gasoline prices in a US city (Gas stations signs of the 1940s) as well as historical prices for crude oil in the US for the time period we will be reviewing in this chapter (EIA public website). Using a Bureau of Labor Statistics website consumer price index inflation multiplier (Padgett 1951), that 20 1/2-cent (USD) gasoline price in 1948 would be about $2.15 in 2017 dollars.

While it may seem too narrow to discuss generally only the US oil and gas climate, this is not the case for the hydraulic fracturing stimulation process. From its inception into modern times, we have observed between 85% and 95% of applications within the borders of southern Mexico through northern Canada, with predominately the US borders alone hosting most of the major history and developments of hydraulic fracturing applications.

To our great collective surprise, as we moved into our sixth decade and into the twenty-first century, we witnessed an unexpected and unprecedented activity increase of hydraulic fracturing applications, similar to a snowball rolling downhill! This came as many of our old and well-known shale "resource-rock" formations were transformed into massive commercial gas and oil plays. In this chapter, we will walk our way through almost seven decades to see what has gradually led our industry to the unexpected new fracture stimulation era we have witnessed during the most recent 15 years.

1.3.2　1947 to 1953: How the Hydraulic Fracturing Stimulation Process (Hydrafrac) Began

1.3.2.1　Era of the Invention of Hydraulic Fracturing

The concept of intentionally breaking down an oil or gas formation using injected fluid to use pressure upon and crack open the rock and then placing solids to keep that crack open was being discussed among engineers with

Stanolind Oil Co. before 1947. However, it was not until early that year that they approached Halliburton Oil Well Cementing Company (HOWCO) at their main offices in Duncan, Oklahoma, with the prospect of signing a secrecy agreement and helping them develop a new concept in well stimulation. Working together, in June of 1947, the first field attempt was staged on one of their wells, the Klegger #1, in Grant Co., a Hugoton formation gas completion in southwest Kansas. At first, the treatment seemed to have failed; however, as there was disagreement as to the need to add a chemical breaker to the 1000 gallons of gelled gasoline carrying approximately 100 pounds of quartz sand, none was added. After approximately a week and some injection of fluid with a breaker chemical, the gel apparently had thinned sufficiently to allow the start of gas production, and they ultimately saw a moderately successful stimulation result.

During the next 12-plus months, several other wells were treated, with a definite successful outcome on almost all. In December of 1948, Stanolind was issued a patent on this new well-stimulation process. For their help in proving this concept, HOWCO was granted a three-year exclusive license to commercialize the hydraulic fracturing process using the Stanolind Oil Co. trade name "Hydrafrac" (although Stanolind could use any pumping company on their own wells). This new license was first used in March of 1949 on an oil well in Velma, Oklahoma (Figure 1.2). Strangely enough, the Velma oil field in the 1920s was the original drawing card that caused Earl P. Halliburton to start his company (HOWCO) in Duncan, Oklahoma, the closest town to the Velma oil fields with railroad service. Note: Stanolind Oil Co. was absorbed into Pan American Petroleum Corp., which was blended into Amoco Oil Co. in approximately 1970. By this time, the Hydrafrac patent had expired (although Amoco's research personnel could still claim "fatherhood" to the process). In the late 1980s, Amoco merged with British Petroleum (BP) to become BP Amoco, later renaming to just BP.

1.3.2.2 Another Important Commercialization of 1948

Another patent issued in 1948 was also a major boon to the production of oil and gas—the use of shaped charge explosives to "shoot" perforations in

FIGURE 1.2
Photos from March 1949 on location for the first commercial application of the Hydrafrac process. (From Halliburton.)

well casings to allow communication with formations in cemented wells. This quickly proved to be a less expensive perforating method (arguments about "better" still exist) than bullet gun perforating and much less costly than hydrajet abrasive perforating, which required at least a workover rig. It is impossible to fully comprehend how much of the success of hydraulic fracturing in cemented wells might have been boosted by the development of shaped charge perforating. Of course, shaped charges also would occasionally create some problems (or at least limitations) to fracturing applications, but that is a small percentage of the time.

1.3.3 Mid-1950s to Early 1960s: The Beginning of Fracturing Applications

1.3.3.1 Commercialization of Hydrafrac Broadens to Other Service Companies Allowed to License

By the start of 1950, several hundred fracturing treatments had been pumped by HOWCO, with a 73% success rate as judged by their customers (Padgett 1951). Later that same year, hydraulic fracturing was used outside the US for the first time in the Cardium oil field in central Alberta, Canada.

Following the end of their initial three-year exclusive license (at the start of 1952), HOWCO renewed their license, and two other companies were also granted a license. By midyear, the total number of Hydrafrac treatments that had been performed crossed the 20,000 threshold (National Petroleum Council 1967). During 1953, the job count averaged about 2300 per month, with other pumping service companies such as Dowell, The Western Co., and Cardinal Chemical being some of the other major participants.

Throughout the early 1950s, jobs were usually small (~1500–4000 gallons of injected fluid), and only about 0.50 lbm of proppant/gal (ppg) was placed, typically at rates of 3–4 bbl/min. A 4000-gallon treatment or injection rates of more than 4 bbl/min (where two pump trucks would be needed) was considered extravagant by many. These small treatments could usually be no more than "damage bypass" treatments to better communicate the formation to the wellbore. Many were in open hole, with a greater percentage in cemented wells as time passed. Figure 1.3 shows a typical truck (Hassebroek et al. 1954) used for mixing proppants (usually washed and graded sand) into the fracturing fluid being fed to the pressure pumping unit(s). You won't need to look closely for instrumentation, as there was none.

1.3.3.2 1955 Proved to Be the Peak Year during the Twentieth Century

With almost 45,000 fracturing jobs recorded, treatment size now averaged approximately 7000 gallons, placing approximately 1 ppg and mostly using a thin crude oil to begin the fracture growth and a more viscous crude to carry proppant. There were just under a monthly average of 2800 rigs, which drilled approximately 57,000 wells that year. These totaled about 212 million ft of hole

FIGURE 1.3
Typical field truck used in the mid- to late 1950s for proportioning proppant into the fracturing fluid. (From Hassebroek, W.E., Stegelman, A., Westbrook, S.S.S. 1954. Progress in Sand-oil Fracturing Treatments. *American Petroleum Institute*, New York, API-54-212.)

drilled with an average depth of ~3700 ft. However, with many wells being "fraced" in more than one interval and "refracs" also occurring, the number of wells treated was less than the number of jobs. Another early development was the use of granular diverting agents to terminate growth of a fracture and force formation breakdown in another area of an open-hole completion or in another zone if it was cased and multizone perforated. Therefore, some of the fracturing jobs represented multiple wellbore locations being fracture stimulated during one job (Clark and Fast 1952).

A typical pressure recording during a fracturing treatment in the mid- to late 1950s was a simple circular plot on a Marten-Decker recorder (Figure 1.4), where the pen would move in/out as the chart paper was rotated (Hassebroek et al. 1954). By the mid-1960s, this recording was improved to use of three pens so the pumping rate, sand concentration, and wellhead pressure could be simultaneously plotted on the chart; however, as treatment sizes grew larger, the data could not all fit on one piece of circular chart paper.

1.3.3.3 A Common Belief Was That Hydraulic Fractures Were Primarily Horizontal Pancakes

Many of the early papers even used the term "hydraulic lift" for the process that occurred, and for the more shallow wells, that might have often been the case. The applications in open holes, where cement does not allow the operator to specifically dictate where hydraulic pressure is applied, can also

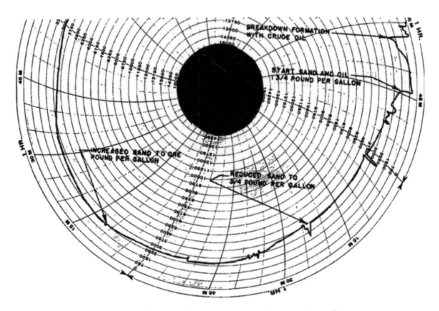

—**Pressure Recording Showing Lowering of Pressure during Treatment**

FIGURE 1.4
Marten-Decker circular pressure recording of data while pumping an early-day fracturing treatment. (From Hassebroek, W.E., Stegelman, A., Westbrook, S.S.S. 1954. Progress in Sand-oil Fracturing Treatments. *American Petroleum Institute*, New York, API-54-212.)

enhance the effect of weak rock boundaries, some of which would even be the location of drilling wash-outs that could also increase this chance in the immediate wellbore area. Some early efforts to understand the process reported use of a downward-facing camera with a mirror to examine open-hole fracturing results in relatively shallow wells (Clark and Fast 1952).

While a few early technical papers tried to change the commonly "accepted" concepts about the behavior of hydraulic fracture growth, only a small number of well operators were concerned with anything technical.

1.3.4 Late 1960s: Expansion of Basic Knowledge of Downhole Fracturing Events

As we moved through the final years of the 1960s, the oilfield began to see more widespread science being introduced as well as significant field experimentation. Among the more important concepts were some that specifically addressed how we should commonly expect vertical fracture planes, particularly for depths greater than ~2000 ft. Additionally, the experience of multiple fractures forming at the wellbore, where nature has been greatly disturbed by the act of drilling a wellbore, should not be surprising, but the likelihood of maintaining far-field growth of them is less than a single dominant fracture (or only a very few) continuing to be developed.

Physical studies in open-hole completions examined before and after hydraulic fracturing, such as described by Anderson and Stahl (1967) using inflatable impression packers or borehole acoustic logging televiewer reported by Zemanek et al. (1969) further illustrated multiple inclined fractures at the wellbore. The (essentially vertical) fractures were common (and even some horizontal fractures) but seldom the only fracture inclination observed, but as a T-shaped fracture. Most of this field experimental evidence of multiple fracture planes being dominant was ignored since the oilfield was moving toward cemented wellbores. Operators wanted to assume that cement would reduce, if not eliminate, nonideal fracturing behavior, and if/when cementing quality was good, this was an acceptable assumption for the majority of applications.

1.3.4.1 Study of Fractures Using "Expanding Open-Hole Impression Packers"

In the late 1950s and into the mid-1960s, there was some use of "balloon packers" made from heavy fiber-reinforced tubes (Anderson and Stahl 1967). On location before running into a fractured open-hole section, they were hand painted with a layer of curable rubber and run down to a hydraulically fractured open-hole section of the pay zone (on tubing). Once at the desired depth, they would be hydraulically inflated and then held in that condition for a few hours while the wellbore impression was formed onto the curing rubber. After curing time, the packer was deflated and retrieved, laid out on location, and photographed. It was also closely examined before being disturbed or loaded to take to the laboratory. It was necessary to run multiple packers for zones longer than about 20 ft, and in a few cases, a prefracturing impression packer would be run. Since this prefrac process was time consuming and would delay the Hydrafrac operation, it was only performed on a few of the 20-plus total wells where this investigative tool was reported as being successfully used.

Some (likely half or less) of that photographic evidence (and many negatives) still exists; however, many of the photos were of poor quality originally. Only very limited portions of these data were published at that time by Anderson and Stahl (1967) and a few others. An extensive review of the remaining photos and negatives was performed in recent times, and a portion of that work was republished (McDaniel 2010). Many of the more interesting parts of the tube pieces, with interesting propped fractures showing clearly, were sawed to short (12–20-inch) pieces, and were often used in seminars and classrooms. However, by the early 1980s, essentially all of those samples had disappeared. Many oilfield personnel were first convinced we almost always created vertical fractures by seeing and touching the sand in the molds of the fractures.

In the 14-plus wells that were included in the impression packer data summarized by Anderson and Stahl, there was only one record of finding

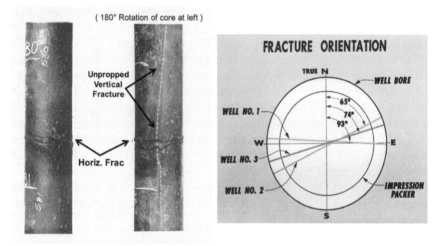

FIGURE 1.5
(Left) The 580-ft propped horizontal fracture. (Right) The vertical fracture strikes observed in three close offset wells. (From Anderson, T.O., Stahl, E.J. 1967. *Journal of Petroleum Technology* 19(2), 261–267. McDaniel, B. W. January 1, 2010. *Archives of Wellbore Impression Data From Openhole Vertical Well Fracs in the Late 1960s.* Society of Petroleum Engineers. 10–12 February, Lafayette, Louisiana, USA. doi:10.2118/128071-MS.)

only a propped horizontal fracture, which was in an oil zone only about 580 ft deep, and close examination of those photos showed a nonpropped, paper-thin vertical fracture both above and below the fat, propped horizontal fracture image (left photo in Figure 1.5). The other well depths investigated varied from approximately 1200 to 3000 ft. Few other horizontal fracture impressions were observed, but those all had tall vertical propped hydraulic fractures, usually above and below the one horizontal impression within the zone. Also reported by Anderson and Stahl were efforts in one field to identify the strike of the vertical fractures. The photo on the right of Figure 1.5 shows the strike of three close offset wells, as per the resulting impressions observed after Hydrafrac treatments on each.

Most commonly, the vertical fractures angled through the wellbore at a slight angle (2–8°), with two cases observed at a 15–20° angle. More than half of the total wells examined had multiple fracture planes intersecting the wellbore. Figure 1.6 shows two cases from those impression packer studies. The left photo shows a front/back view with single fracture just below foot marker 39 and dual fractures 1 ft lower. The right photo shows a fracture at a sharp angle (dip) with about 3 ft of the fracture height penetrating the wellbore.

1.3.4.2 Organization of the Petroleum Exporting Countries Was Formed

In September of 1960, an event occurred in Baghdad, Iraq, that would have a greater impact on the use and application of hydraulic fracturing than any

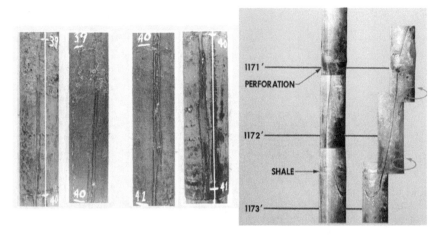

FIGURE 1.6
Two different cases of impression packer testing results. (From Anderson, T.O., Stahl, E.J. 1967. *Journal of Petroleum Technology* 19(2), 261–267. McDaniel, B. W. January 1, 2010. *Archives of Wellbore Impression Data From Openhole Vertical Well Fracs in the Late 1960s*. Society of Petroleum Engineers. 10–12 February, Lafayette, Louisiana, USA. doi:10.2118/128071-MS.)

other single event since its inception and perhaps even currently—OPEC was formed. Headquartered in Vienna, Austria, since 1965 (originally in Geneva, Switzerland), OPEC is an intergovernmental organization founded by its first five members: Iran, Iraq, Kuwait, Saudi Arabia, and Venezuela. However, it was almost a decade before hydraulic fracturing applications first felt major effects from this organization.

In 2017, OPEC was an organization of 14 oil-exporting member nations, specifically excluding the US, all of Europe, Russia, China, Australia, and all of South America other than Venezuela and Ecuador.

1.3.4.3 Hydrafrac Update through 1963

By 1964, more than 400,000 Hydrafrac treatments had been pumped as most oilfield operator companies accepted the process. However, many still believed that pancake-type fractures and lower sand concentrations could result in a partial monolayer with higher conductivity than packed fractures, helping keep the interest in low-proppant concentrations alive. The graph on the left of Figure 1.7 is from a 1964 *Journal of Petroleum Technology* (JPT) article by Hassebroek and Waters (1964) and shows that sand and fluid volumes were steadily increasing, but jobs were still small and sand concentrations stayed relatively constant. The graph on the right of Figure 1.7 is a chart showing how oil-based fluids were gradually superseded by water-based fluids where the water was often gelled. By 1963, the average pumping rates had increased to about 18 bbl/min and the typical hydraulic horsepower used was about 1200 HHP, requiring two to three pumping units.

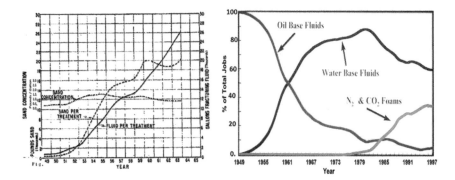

FIGURE 1.7
(Left) Average fracturing treatment sizes in the US from 1949 to mid-1963. (From Hassebroek, W.E., Waters, A.B. 1964. Advancements Through 15 Years of Fracturing. SPE 801. *Journal of Petroleum Technology*, 760–764.) (Right) Base fluids used for fracturing from 1949–1997. (From Halliburton.)

Figure 1.8 shows the data from a National Petroleum Council study titled *Impact of New Technology on the Petroleum Industry 1946–1966*. This 346-page public document (National Petroleum Council 1967), with hundreds of total contributors assisting the various committees, provides an oil production progress overview of oilfield advancements during the two-decade era after World War II. Included in the conclusion (Chapter 1, "Conclusions and Summary; Sub-Section 2—Conclusions, Part C: Productive Capacity") is the following comment about improved productive capacity: "The two most

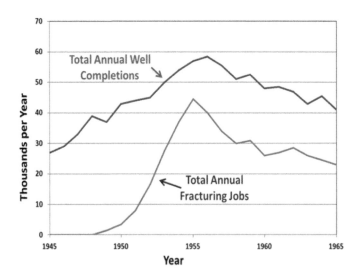

FIGURE 1.8
Number of well completions and fracturing jobs per year. (After National Petroleum Council. 1967. Impact of New Technology on the U.S. Petroleum Industry 1946–1965. Library of Congress Catalog Card Number: 67-31533.)

significant improvements of technology are water injection (flooding) and hydraulic fracturing."

1.3.5 Late 1960s to Mid-1970s

1.3.5.1 Project GasBuggy (1967–1973 Underground Nuclear Experiments)

Project GasBuggy was the first in a series of US Atomic Energy Commission's (AEC) downhole nuclear detonations to stimulate production from low-productivity gas reservoirs. The first of three total attempts, this was one of many projects funded under the AEC Project Operation Plowshare (begun in 1957 and terminated at the end of the AEC's fiscal year 1975) that attempted to develop peacetime applications of atomic energy. For many years before hydraulic fracturing was invented, nitroglycerin had been used to "shoot" open-hole sections of oil-bearing formations, rubblizing rock and creating a lower resistance path for oil to flow though the wellbore. This led to the idea that the power of nuclear detonations could provide a massive zone of "nuclear fracturing" rubble to super-stimulate gas production, which could eventually become a commercially viable oil and gas well stimulation process.

In 1967, after offsetting three older low-production natural gas wells, the project team drilled an abnormally large wellbore for Project GasBuggy to a depth of 4240 ft and cementing casing to approximately 4000 ft, leaving an open-hole zone in the gas-bearing zones to host the nuclear explosion, as illustrated in Figure 1.9. This well location, shown in the photo, was in the Carson National Forest, approximately 90 miles northwest of Santa Fe, New Mexico. One of the offset wells that contained various sensing devices for monitoring is shown in the background.

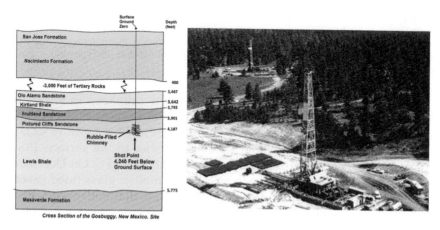

FIGURE 1.9

(Left) Wellbore schematic; (right) photo of wellsite of Project GasBuggy. (From *Project Gasbuggy Wikipedia*, accessed 6/27/2017. https://www.google.com/search?as_st=y&tbm=isch&as_ q=project+gasbuggy&as_epq=&as_oq=&as_eq=&imgsz=&imgar=&imgc=&imgcolor=&i mgtype=&cr=&as_sitesearch=&safe=images&as_filetype=&as_rights=.)

FIGURE 1.10
Gas Buggy nuclear device while the crew prepares to install it into the large wellbore. (From *Project Gasbuggy Wikipedia, accessed 6/27/2017.* https://www.google.com/search?as_st=y&tbm=isch&as_q=project+gasbuggy&as_epq=&as_oq=&as_eq=&imgsz=&imgar=&imgc=&imgcolor=&imgtype=&cr=&as_sitesearch=&safe=images&as_filetype=&as_rights=. Google free to use images.)

Figure 1.10 shows the device while scientists prepared to lower the 13-ft-tall by 18-in.-diameter nuclear warhead into the well. The plan was to detonate the experimental 29-kiloton (kt) device at a depth of 4227 ft after the wellbore was backfilled above the device to the surface. (As a frame of reference, the Hiroshima bomb was about 15 kt.) On December 10, 1967, the device was detonated, creating a molten glass-lined cavern approximately 160 ft in diameter and 333 ft tall that collapsed within seconds. Subsequent measurements indicated fractures extended more than 200 ft in all directions. The well was untouched for two months for radiation decay before the wellbore was cleaned and then flow-tested on flare for 30 days, three times during 1968. Gas production was improved compared to earlier offset wells but was disappointing and had radioactive contamination.

Later, two subsequent nuclear explosion fracturing experiments were conducted in western Colorado in an effort to refine the technique: Project Rulison in 1969 with a 50-kt charge and Project Rio Blanco in 1973. In both cases, the gas radioactivity was still observed as too high, and in the last case, a triple 30-kt blast created rubble chimney structures that were disappointing because the rubble was excessively fused and there was concern many of the extending fractures might have impermeable faces. Flow testing to predict recoverable gas indicated that total payback on these projects would only be between 18% and 45%. Additionally, public concern about radioactivity was growing, and returns were not expected to become economic. There were three additional project tests started in Alaska, but the Operation Plowshare Project ended without any additional wells seeing nuclear devices installed.

When these nuclear fracturing projects initially appeared, it was widely speculated that they may at least partially reduce the market for hydraulic fracturing. However, the actual outcome only enhanced the need to further develop the 30-year-old fracturing stimulation method.

1.3.5.2 When Perforated Wells Won't Break Down to Allow Fracturing

This topic is included to illustrate how little most who were operating in the oilfield understood about rocks, perforating, and other downhole rock conditions. Even for those with a higher understanding of rock behaviors, there was still the issue that most well conditions did not truly match what was envisioned to be the formations' in situ conditions. When "fracturing experts" received field phone calls in hopes there was some magic answer for overcoming this breakdown problem, the caller was seldom able to furnish much real data about the well condition, actual formation, actual cementing specifics, perforating gun/charge data, or sometimes even the actual fluid at the perfs at the time the breakdown was attempted. The caller might or might not know whether this was a common problem to the field or the formation. The stated depth of perfs and formations would usually be available, but very often little else other than that the answer was needed ASAP, if not instantly. Few of the mainstream oilfield engineers of the 1970s era were very knowledgeable about any technical issues of hydraulic fracturing.

1.3.5.3 Bringing Rock Mechanics to Fracturing Technology

During the early 1970s, we witnessed most major oilwell operators and the larger pumping service companies either creating internal rock mechanics laboratories or greatly expanding these labs. However, for many well operator companies, this laboratory was primarily focused on improving the drilling processes, since drilling success was their mainstay. However, as witnessed in publications, we did see a few majors (Mobil, Exxon, and Amoco, for example), and the smaller labs of the pumping services providers (such as

HOWCO and Dowell) were often focused more on the rock mechanics of fracture stimulation. The technical papers seemed to indicate we should assume vertical fracture orientation for all except the most shallow wells, and with cemented wells, the possibility of T-shaped fractures should become less prevalent. This was not always the common thinking of the many smaller operators who were drilling wells.

Unfortunately, the problem of initial formation breakdown is still a challenge that occurs in modern operations where inclined fractures are common. The wellbore has altered the natural stress condition, cementing conditions are not always the same, and perforations do not always penetrate the exact same rock. It would be the 1980s before the oilfield increased its focus on how creating perforations with great physical force on a small portion of the wellbore would often introduce anomalies that made near-wellbore (NWB) fracture mechanics behave much differently than the main body of the rock formation away from the NWB area.

In the 1970s, the potential remedies for initial formation breakdown issues were few. Soaking the rock with 5%–10% hydrochloric (HCl) acid was the most common solution, and it often worked, even though modern-day acid-soluble cement was not used then and rarely did these formations have significant acid solubility. Misplaced perfs were thought to be another fairly common reason, which resulted in trying to break down in a dense shale instead of the intended rock. Sometimes the answer was reperforating using a "name-brand" perforator, and sometimes the use of fresh water was thought to be the culprit. When none of these appeared to be the answer, the last straw would be to rig up, run a hydrajet tool into the well, and cut a slot with abrasive jetting. This approach never seemed to fail but was not usually the answer for the "next well" as a prevention method because of expense and time, since a workover rig was necessary when the drilling rig had been released (there was little to no coiled tubing [CT] available then).

Core studies were few and of limited technology with respect to fracturing stimulation. For example, the first Society of Petroleum Engineers' (SPE) publication where scanning electron microscope (SEM) evaluations of formation rock were used to help assist/improve low perm core fracture stimulation fluid recommendations was not published until 1976 (Simon et al. 1976).

1.3.5.4 Introduction of Handheld Calculators to Oilfield Applications

To those reading this who were born after 1970, the concept of using a slide rule to speed up multiple or intricate calculations probably will not register with them, since few will know what one was (Figure 1.11). However, handheld calculators, and then later preprogrammed and finally programmable handheld calculators, were quite a boon to the oilfield by the dawning of the 1980s—a major boost in efficiency only the old-timers can comprehend currently.

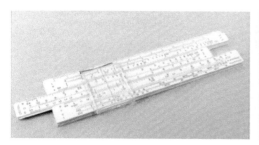

FIGURE 1.11
Typical slide rule used by math and engineering applications up until the early 1970s. (Photos from Wikipedia download.)

1.3.5.5 Birth of Computerized Fracture Simulation Modeling

As one who taught seminars and short technical schools to thousands of operators' personnel for a decade following the early 1970s, the author witnessed firsthand that more than half that group came into those rooms assuming hydraulic fractures were horizontal pancake-shaped cracks, but all left with a belief in vertical fractures.

During the early to mid-1970s, the leading computer technology used was punched card decks, which, again, many readers might not be old enough to comprehend. Oftentimes, trying to run a specific set of well and stimulation input data would fail, and the error in the data or the program itself would have to be resolved even to get initial answers. A mispunched card in the data card deck was always the hoped-for answer, not a programming error. These fracture growth simulators were owned by some major company well operators, a few major service companies, and only a few universities. Additionally, there were multiple schools of thought as to the proper rock mechanics prediction theories to be used in programming hydraulic fracture growth simulations, with few, if any, of these theories having extensive laboratory verification.

However, even when the "home office" might have access to a computer, the day-to-day life of oilfield chemists, managers, and/or engineers was serviced by printed tables, charts, and graphs (plus their slide rules).

1.3.5.6 Arab Oil Embargo from 1973 to 1974

The first major worldwide use of oil supply as a geopolitical tool was during the war between Israel and Syria/Egypt in October 1973. Because the Arab community largely controlled OPEC, it declared an embargo on oil shipments to the United States because it was the primary power supporting Israel. The Arab members extended that embargo to other countries that had also supported the US stance during that military conflict. Although the United States was a major oil-producing country, in 1970, US domestic oil production peaked at just above 10 Mbbl/D (million barrels per day) and oil imports were

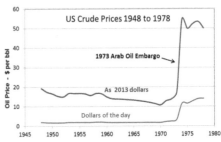

FIGURE 1.12
(Left) In late 1973, lines at gas stations went for blocks, with a limited quantity allowed per car. (From Arab Oil Embargo 1973: Google free to use images https://www.google.com/search?as_st=y&tbm=isch&as_q=arab+oil+embargo+1973&as_epq=&as_oq=&as_eq=&imgsz=&imgar=&imgc=&imgcolor=&imgtype=&cr=&as_sitesearch=&safe=images&as_filetype=&as_rights=.) (Right) Oil prices in the US between 1948 and 1978 (in money of the day and in 2013 dollars). (Data downloaded from Baker Hughes public website.)

approximately 1.2 Mbbl/D. From there, domestic production fell and imports had grown to more than 3.6 Mbbl/D in 1973, and much of this import volume was coming from OPEC member nations.

Before the embargo, US crude oil was approximately USD 3.50 per barrel (approximately $16 in 2013 dollars). Within days after the embargo was announced, the price doubled and was above USD 10 per barrel before the end of the year. With general panic and increasing prices troubling the distribution chains, there were long lines of vehicles at gas stations in every major city in the United States and many other countries (Figure 1.12). This effectively altered the world balance of power in the international climate. OPEC had already been successfully using its "oil power" to its benefit, and following the terms ending its embargo, it emerged even more powerful by 1975 and reduced the worldwide influence of the United States.

1.3.6 Birth of Coalbed Methane Fracturing and Massive Hydraulic Fracturing

The initial application of these two technologies was not widely practiced until several years later, but both are mentioned here while we move chronologically, as both first occurred in the second quarter of 1974.

1.3.6.1 Coalbed Methane Fracturing

There was some sporadic fracturing of coal seams to enhance degassing before the case mentioned here, but the Inland Steel coal mine in Sesser, Illinois, was the first case in the world where oilfield technology personnel would get to enter an active coal mine to visit an area where a wing of the propped hydraulic fracture was exposed. This well had been drilled through the coal as a methane vent had cemented casing from the top of the coal to the surface. An open-hole section was then drilled into the coal, and the coal zone was hydraulic fracture–stimulated, with the intent that it could be examined after it had been mined through. This was an ideal case because the coal seam was about 9 ft thick in this area and was being mined in a pillar-and-post method; in this case, the tunnel's width was approximately 12–13 ft and unmined "pillars" were approximately 30–40 ft square.

About the first of May, one wing of the hydraulic fracture had been mined through with propped fracture visibility on five walls of the pillars. Because the overlying sandy shale formation above the coal would soften and weaken if exposed to moisture from the humid mine air, about 10–12 in. of "roof coal" was left unmined until there was time to install girder reinforcements of the shaft in an area. The following week, it was scheduled to move active mining away from the fracture intersected zone to allow for scientific inspection by fracturing technology personnel. The inspection was led by a coal mine operations manager and a mine geologist, while two Halliburton personnel (an engineer from research and the local Halliburton camp district engineer) were transported to the fractured area more than a mile away from the mine shaft since Halliburton Research was involved in a new coal fracturing–related multicompany research program. Using special "nonsparking cameras," each area was inspected and photographed where propped fractures of approximately 1/8–3/16 in. were present on both the left and right faces of two parallel tunnel areas and only on the left face of the third parallel tunnel area because the borehole that was fractured was 80–100 ft to the left of the initial inspection area. In all areas, the fracture was proppant filled to the roof coal. By standing on the bumper of the small transport vehicle, the roof coal in the first area was chipped back near the tunnel's center until it reached the overlying rock, as all were curious to see if the proppant was up there also. Proppant was present through all the coal, but only a narrow closed crack with no proppant was visible once the overlying rock was chipped into and exposed for several inches of crack length.

This chronology will revisit coalbed methane (CBM) fracturing in approximately 10 years. Figure 1.13 shows a later example of observing fractures in the face of coal blocks (pillars) where multiple fractures were evident parallel to the face cleats of the coal on both sides of the wellbore. This example is included here to better visualize the pillar-post coal mining method.

1.3.6.2 Birth of Massive Hydraulic Fracturing

The first occurrence of any fracture stimulation operation placing more than one million pounds of proppant into one zone during one pumping operation was in late May 1974. The well was an Amoco Oil and Gas Co. completion in the Muddy J sand about 8000 ft deep in the Denver Julesburg (DJ) a few miles north of Brighton, Colorado. The fracturing fluid was a water-external emulsion fluid where the water phase was gelled using 40 pounds per thousand gallons (lb/Mgal) of guar gelling agent, and the hydrocarbon phase was a 58–60° condensate from nearby wells. This was a time in the oilfield when there was minimal

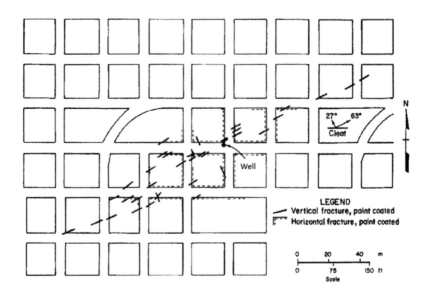

FIGURE 1.13
Example overhead view of pillar-post mining case that had mineback observations of propped and nonpropped hydraulic fractures at many places on both sides of the wellbore where fractures were located in walls of coal blocks that serve as pillars. (From Halliburton *CBM Handbook* 1991.)

onsite sand storage. Approximately 35% of the sand was delivered to the fluid mixing blender initially using multiple body-load trailers. When these were empty, they were moved away from the blender and the remaining proppant sand was delivered in an "ant train" of dump-style trucks (Figure 1.14). The many smaller dump-style trucks were reloaded once or twice at one of the two upright storage bins approximately 500 ft away from the blender.

Prior to this historic fracturing treatment, there had actually been two earlier well completions where the operator would have attempted this first "massive hydraulic fracture" (MHF) treatment, but rainy weather delayed them since more than 90% of the proppant needed required movement on location to execute such an operation. Generally, approximately 300,000–500,000 lbm total had been the previous standard completion in similar wells in this field (Fast et al. 1977). More widespread and much larger applications of the MHF method of treatments will be discussed later.

1.3.7 Late 1970s to Early 1980s: Proliferation of Knowledge and Application of Fracturing

1.3.7.1 Introduction of "Frac Vans" to the Oilfield

For years, the oilfield mostly watched pressure gauges set on the back of a pickup tailgate or under tents or awnings when necessary. With advances in

FIGURE 1.14
Aerial photo of the location when the first million-pound treatment was approximately 75% complete, on May 1974. (From Halliburton.)

analog computing power, fracture treatment field operations began to see that visual data displays were starting to be installed inside small vans to protect electronics and have a place for the lead operator personnel to get out of the weather to better observe the live data for pressure, pump rates, and fluid/sand mixing while the job progressed. As these frac vans evolved, they were housed in larger trucks and began to incorporate greater computing power and digital memory, which will be discussed later.

1.3.7.2 Introduction of the "Pillar Fracturing" Technique

In the early 1970s, fracture acidizing was the first stimulation process that used fluid density control to introduce a pillar-support concept within acidized fractures. In late 1974, a new patent was issued for use of viscous gels supporting sand-laden fluid alternating with sand-free fluid used in a pillar fracturing concept with hydraulic fracturing (Tinsley 1974). One of the early technical papers to report on extensive field applications was work done on 24 wells during 1976 in west Texas (Pugh et al. 1978). The process saw many applications through the next several years and has again been revived by several service companies in very recent years (i.e., post-2010).

1.3.7.3 Gas Research Institute Founded

The Gas Research Institute (GRI) was founded in 1976 in response to the Federal Power Commission (FPC) encouraging increased gas research and development (R&D). The GRI administered research funding provided by a surcharge on shipments of natural gas sold by the interstate pipelines. (Note: At its peak in 1994, the GRI administered funds in excess of USD 212 million.) For the next two decades, the GRI would play an extensive and important part in fast-forwarding the industry's increased knowledge of tight gas formations (mostly sandstones) and the technologies necessary to commercially produce gas. Most of this would be centered on our better understanding of "how/when/where we should apply hydraulic fracturing, and what is occurring inside the rocks while we do this."

One of the first efforts was the overall GRI Tight Gas Sands Research Project that was started in 1982 (O'Shea and Murphy 1982). First, the GRI drafted comprehensive programs intended to increase the national supply of natural gas from unconventional sources, and low-permeability (tight) gas sands were the initial unconventional source targeted. The GRI proceeded initially with six project types to be completed by 1987: Resource Identification, Formation Evaluation, Fluids and Proppants Investigations, Fracture Design, Reservoir Modeling, and Staged Field Experiments (SFE) with technology transfer. However, in 1983, the GRI added the Multiwell Experiment (MWX) that extended through 1986 to this list.

Some of this work under GRI guidance will be discussed later in this chronology.

1.3.7.4 Development of New Fracturing Fluids

Viscous gels, crosslinked gels, gelled water-external emulsions, and foams (both nitrogen based and carbon dioxide based) were implemented as the fracturing fluids most often used. Fracturing applications quickly turned away from crude oil as a fracturing fluid. We saw in Figure 1.7 that by 1961, half the fracture treatments pumped were using water-based fluids, and by 1984, approximately 11% of treatments used any type of oil-based fluids, and these usually were gelled diesel or lighter hydrocarbons. Also that year, we saw there were as many foamed fluid treatments as hydrocarbon. Generally, the foamed fluids or a hydrocarbon base (gelled diesel, kerosene, or condensate) were used only where formations were judged too water sensitive to apply any of the numerous new water-based gel systems, even when sodium chloride (NaCl) or potassium chloride (KCl) was added as a temporary clay stabilizer. Additionally, a water external emulsion fluid was evaluated for several years, since the water phase would typically be approximately 1/3 of the fluid. By gelling this phase, the fluid viscosities were double or triple that of the gelled fluid, with generally diesel or lighter, up to more than 60° American Petroleum Institute (API) condensate, that would serve as the emulsion system's internal phase. This water external emulsion fluid was the fluid system used for the first million-pound fracturing treatment in 1974 mentioned earlier, but by the early 1980s, this fluid was rarely being used.

As we began to need to fracture-stimulate deeper and hotter formations, the search for fluids that could open fractures and transport high proppant concentrations led to many new fluid systems where the base gel was crosslinked to produce 5- to 20-fold higher fluid viscosities in the fracture than simple linear gel systems could produce that would still have acceptable pipe friction properties and could still be degraded adequately for good postjob fracture cleanup.

In reservoir engineering of fracture-stimulated wells, evaluating the effect of fracturing on initial and ultimate production was an important part of designing fracturing treatments. Evaluating the economics of the fracturing treatment is the drive for predicting the length, width, and conductivity of a fracture, which dictates the range of pump rates to be used, the amount of proppant, proppant concentration, and so on.

Initial estimation of production improvement focused on representing a fracture as a skin factor; therefore, a fractured well would be represented as a well with negative skin. This could be viewed as the well represented by a wellbore with a larger radius; thus, the fluid flow would be essentially radial. This concept is acceptable when the formation permeability is fairly high and the fracture length is fairly short, causing the fractured well to reach pseudoradial flow in a fairly short time. This representation also simplifies the calculations of the production rate, and many of the early fracturing treatments were successful with small treatments since NWB damage could be bypassed.

It was quickly realized by the 1970s, in the majority of cases where fracture stimulation was applied, that the formation permeability was low, which was the main reason a well would be hydraulically fractured, and the job sizes were drastically increased. Consequently, a long fracture was considered desirable and, accordingly, treatments that were designed would mean that the skin factor representation would not be accurate. Development of steady and pseudo–steady state modeling tools followed. It was understood that fluid flow around a fractured well would reach the steady state in a short time, which required the development of more sophisticated solutions, both analytical and numerical. Those solutions represent the phenomenon around the fracture more accurately and realistically.

Using the equivalent skin factor from the beginning would cause a significant underestimation of productivity. The error becomes larger as the formation permeability decreases. Numerical simulation of well performance using the rigorous fractured well behavior versus using an equivalent skin factor to represent the fracture clearly illustrated that concept. The error in predicting production could lead to false financial estimates that, in turn, could lead to erroneous reservoir management decisions. More detailed and technical discussion may be found in *SPE Monograph*, Volume 23, Chapter 11.

1.3.7.5 Expansion of Massive Hydraulic Fracturing

In the late 1970s and especially during the 1980s, the industry saw the proliferation of large fracturing treatments placing more than a million pounds of sand in single-stage treatments into low-permeability formations, mostly gas-bearing sands. Natural gas was in high demand, so prices were high, at least through 1984. As previously mentioned, the case where the million-pound mark was first passed was in May 1974, with two other treatments in this field of similar size the following year; however, it was 1978–1980 before this trend started to be more common in very many tight sands in the United States (Figure 1.15).

In some of the deeper formations where sand would experience severe crushing, higher strength man-made proppants would be pumped as tail-ins. In a small percentage of the wells, high-strength proppants would be used throughout. The first treatment that surpassed more than 200,000 lbm pumped in one stage was in Bakersfield, California, on a deep Tenneco Oil Co. well where 500,000 lbm of 20/40 bauxite proppant was placed in one pumping operation, a record that stood for more than four years. However, it was in the mid-1970s when Dr. Claude Cooke (Exxon) was among the first researchers to show the industry that we were damaging our proppant pack conductivity with our commonly used gelling agents, even at high formation temperatures (Cooke 1975).

FIGURE 1.15
MHF at 13,000 psi wellhead pressure in south Texas, 1978, placing approximately two million pounds of proppant. (From Halliburton.)

1.3.8 Landmark Concepts That Forever Changed Fracture Stimulation Design and Modeling

1.3.8.1 Introduction of Theories of Fracture Extension "Net Pressure" and "Minifrac" to Identify Net Pressure and Fluid Leakoff (Ken Nolte and Mike Smith)

Starting about 1977, there was extensive field research work coordinated for Amoco Research by Dr. Ken Nolte and Dr. Mike Smith, with Bill Miller overseeing much of the field work. They set out to study and better understand observed pressures during hydraulic fracturing treatments. At the annual SPE convention in the fall of 1979, Nolte and Smith presented their conclusions in two landmark papers: SPE 8297 (Nolte and Smith 1979) and SPE 8341 (Nolte 1979) (SPE 8297 was later slightly revised and published (Nolte and Smith 1981) in the JPT). They presented the concept of tracking the "net pressure above closure pressure" (Figure 1.16, left), with net pressure determined before the stimulation using the minifrac test and its analysis (Figure 1.16, right) by using a short fracture injection test prior to the main treatment to allow the measurement of pressure when a fracture had just closed and also a method to estimate the fluid leakoff coefficient while the fracture was closing. In tandem, these are the basis for fracturing net pressure analysis and its concepts. Within a few years, these concepts had drastically revised the way the oilfield viewed and interpreted hydraulic fracturing events and, with only small refinements, are accepted even today.

With these papers, the application of hydraulic fracture stimulation and its analysis, design, and monitoring were soon to become several magnitudes more sophisticated, and thereby more useful. Certainly more scientific! Probably only a very few readers of this text were active in the oilfield of the pre-1980s, and even if so, were unlikely to have been exposed to the

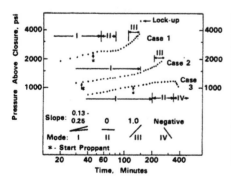
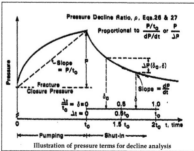

FIGURE 1.16
(Left) The four modes of pressure behavior, Nolte and Smith, 1979; (right) fracture pressure decline analysis of Nolte, 1979.

field data gathering Nolte and Smith undertook to generate their database. Therefore, we will discuss this more extensively. It was unique for its time.

Unfortunately, the general pressure data available for study during and after typical fracturing treatments of the late 1970s, or even as the 1980s began, were not of the exactness or the quality that allowed close or detailed scientific analysis. Figure 1.4 illustrated the original circular job data chart of the 1950s (later improved by a three-pin circular chart), which by 1970 had mostly been replaced by multipen (different ink colors) paper strip chart recordings that included all job data on a single (sometimes long) strip. Typically, this would be a continuous 4- to 5-in.-wide strip that would have pressure data spanning 0–5000, 10,000, or 15,000 psi, depending on expected maximum wellhead pressure. Quality of these data varied with chart speed as well as maximum pressure span, and since rarely would anyone be using the data for intricate postjob calculation (pre-"Nolte-Smith" concepts), the strip speed was generally too slow to identify much more than average readings for 1–2 minutes at best. Digital data, where exact, fractional-second data could be stored for analysis, were not yet in the mainstream fracturing service sector.

To be able to develop a valid data set for minifrac analysis or to accurately observe net pressure, which led to the relationships of net pressure observations, required preplanning for a service provider to provide (normally at extra cost) high-resolution digital gauges, since more than 90% of fracturing field equipment in the early 1980s was using analog gauges of only moderate resolution.

Using standard low-resolution gauges was *not* how the original database of Nolte and Smith was gathered. Their "data jobs" required higher cost, high-resolution (analog) gauges, and the marketplace did not yet have rugged versions usable for everyday pumping service equipment, nor any standard installation digital pressure gauges. Additionally, accurate calculations of pipe frictions were often questionable, so Nolte and Smith's data jobs were pumped down an open-end treatment tubing string so annular pressure could be monitored, or, if pumped down casing, then a 1-in. string of pipe was installed to alternatively provide a "dead string" for pressure data with no fluid friction effects. Either of these methods would give surface pressure data not exposed to pipe friction of any fluids pumped. To obtain pressure-time data recordings of the quality they needed for analysis, a large x/y plotter (typically ~16 × 16 in.) was brought to location, and its pressure recording was scaled with a zero y-axis value to be just below the pressure expected for fracture extension (net pressure). This usually allowed 12–14 in. for recording typically a 2000–3000 psi span of total pressure variations vs. approximately 5 in. for a 10,000 psi scale, as normally recorded on a conventional strip chart in that time period. Also, these unique data were held proprietary to Amoco; if on location, you could view the data being recorded, but only the Amoco research team would have the database from the 100+ wells for analysis and review.

During the early 1980s, for wells where it was desired to observe "net pressure" (or even to accurately calculate postjob), the necessary data recording conditions were not always a possibility, even if the well operator was willing to cover the added costs to obtain such data. The oilfield had new, valuable, practical theories to use but was not yet able to commonly apply them to most fracture stimulation treatments of that timeframe. By the early 1980s, the oilfield was beginning to see more common use of frac vans, but digital pressure transducers were rare in the oilfield at this time. This challenge will be discussed later.

1.3.8.2 The Gas Research Institute Tight Gas Sands Project from 1982–1987

As mentioned previously (O'Shea and Murphy 1982) regarding this GRI project, the GRI proceeded initially with six project types to be completed by 1987: Resource Identification, Formation Evaluation, Fluids and Proppants Investigations, Fracture Design, Reservoir Modeling, and Staged Field Experiments with technology transfer. However, in 1983, GRI added to this list the Multiwell Experiment that initially extended through 1986.

The first of this project work that began to be published were data generated from coring several wells in areas chosen during resource identification. Some of these were based on several "cooperative wells" where operators contributed all well analysis data and stimulation data to the project. The wells selected as Staged Field Project wells were offsets to cooperative wells and where additional data were gathered at the expense of GRI to provide an unusually complete database. The main focus of the research was to help improve the general understanding of producing tight reservoirs and therefore required a specific focus toward advancing the technology involving hydraulic fracture geometry. This project involved "cooperative" wells where well operators would share all data on wells volunteered to the program and SFE wells, where GRI would finance significant additional data too costly for the well operators to fund on their cooperative wells. Also, these SFE wells would offset some of the cooperative wells to build better database information.

Wells SFE 1, in late 1986, and SFE 3 were both in the Waskom field in east Texas, with completions in the Travis Peak formation below 6000 ft. The SFE 2 well was in the North Appleby Field and completed in the Cotton Valley sand. Figure 1.17 shows the location of the cooperative wells and the three SFE wells. This research was very beneficial to identifying formation properties that could aid the fracture simulation models in predicting fracture height and length.

1.3.8.3 Gas Research Institute Sponsored Fracturing
Study at Rifle, Colorado, in 1983

The GRI/Department of Energy (DOE) Multisite Hydraulic Fracture Diagnostics Project was proposed in 1983 and added to the GRI Tight Gas Sands Project, commonly called Multiwell Experiment or MWX. The first "data" papers began to appear in 1984 (Northrop et al. 1984) after coring three wellbores drilled on the site (Figure 1.18). Work at this site eventually (including a later revisit to this site) led to greater added understanding of

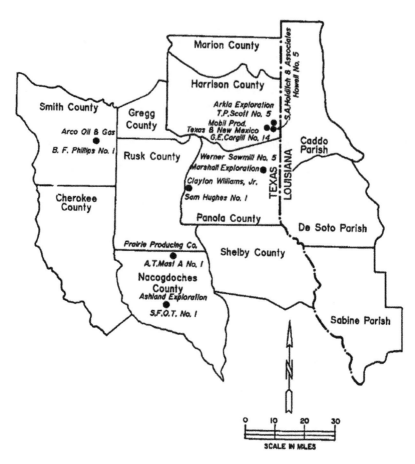

FIGURE 1.17
GRI Tight Gas Sands Project cooperative wells and three SFE wells. (GTI public data.)

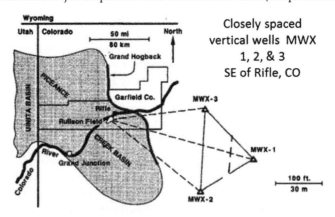

FIGURE 1.18
Well site for the MWX Project. (GTI public data.)

fracture response than probably any other single fracture stimulation test site, just not based on this initial body of work. The early project was conducted at depths of approximately 7000–7100 ft in a tight Mesa Verde sandstone.

The outcome of this initial three-year project and fracture simulations of two zones in this well was disappointing from a production standpoint, apparently suffering from fracturing fluid damage to the natural fracture system and possibly crushed proppant. However, this work at M-site, as it later was often called, would be revisited seven years later by the GRI/DOE Slant Hole Coring Project (SHCP) in 1991–1996, which will be discussed later since the information gained from the SHCP efforts was of major importance.

1.3.8.4 Laboratory Evaluations of More Realistic Conductivity Testing

Spurred by well operator concerns that were initially championed by Dr. Claude Cooke's papers and concerns about gel residue effects on conductivity (Cooke 1975) and high temperature and high-stress effects on fracture conductivities of natural sand (Cooke 1973), the industry began to evaluate both these concerns more closely. Additionally, in 1977, Dr. Cooke published testing showing that sintered bauxite grains could drastically reduce conductivity loss resulting from high temperature and high stress (Cooke 1977). Soon such a proppant was being field tested, confirming the lab data.

In 1981, an American Petroleum Institute committee began the development of new standardized recommended specifications for classifying and testing proppants for fracturing applications, with washed and sieved quartz sand being by far the dominant propping agent materials used, both initially and continuously. One subcommittee first gave us API RP 56 (1983) on proppant general evaluation and classification of fracturing proppants, then API RP 58 relating to gravel pack applications, and another subcommittee continued developing the guide for testing at conditions more similar to what actual packed fractures would encounter to give us API RP 60. Ultimately, these three would be combined years later into one as International Organization for Standardization (ISO) (ISO 13503-1:2003), Part 2.

Guided by the API committee's work (long before the above API RPs were completed), we saw oilfield operators, service companies, and independent testing labs building new types of equipment for testing proppant packs at high temperatures and long testing timeframes (days to months). The initial design concepts for this new type of testing apparatus were a continuation of the initial work by Exxon research, as initially published Dr. Cooke (1975, 1973, 1977). This new methodology/equipment resulted in authorship of numerous SPE technical papers related to this testing in these newly equipped labs (McDaniel 1986; Much et al. 1987; McDaniel 1987; Penny 1987 and Cobb and Farrell 1986). This long-term proppant testing began to revise the mindset of our industry in regards to long-term testing, high temperatures, and high-closure stress. More realistic testing on the behavior of packed fractures brought new comprehensions of realistic fracture conductivity expectations.

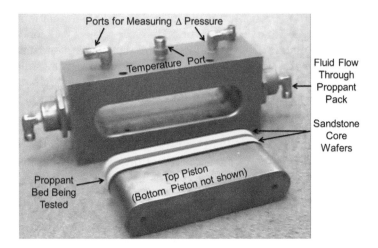

FIGURE 1.19

Example of an API conductivity test cell. (From Halliburton.)

Figure 1.19 shows a representative API conductivity test cell ready to be assembled to place in a load frame, have the plumbing lines attached, then apply closure stress on the proppant pack. The cell can be heated, potentially to more than 300°F, and tests are typically conducted for two to three days and occasionally up to several weeks. Often, several assembled cells are stacked on each other within a single load frame to increase time efficiency of testing multiple proppant types or pack thicknesses. Test fluids used would be deoxygenated and preheated. Gelled fluid effects were also included in many of these tests.

1.3.8.5 Teaching Fracture Stimulation Technology

In the mid-1980s, we saw more universities begin to emphasize broadening their petroleum-related engineering coverage to incorporate a focus on well stimulation practices and technologies. Additionally, several major pumping service companies organized one- to two-week schools for well operator personnel during this timeframe, as many had little original training in techniques and methods now coming into broad use in the oilfield. A few of these became highly accepted and were continued into the late 1990s.

1.3.9 Mid-1980s: Greatest Crash in Oilfield History

1.3.9.1 Glory Years for the United States Oilfield ... Then the Crash

After the rebellion in Iran brought the fall of the Shah and the rise to power of Ayatollah Khomeini in 1979, many areas of the world suddenly faced oil shortages, rapidly driving oil and natural gas prices higher. In only two years, this had driven US rig counts to record highs. The US rig count peaked at

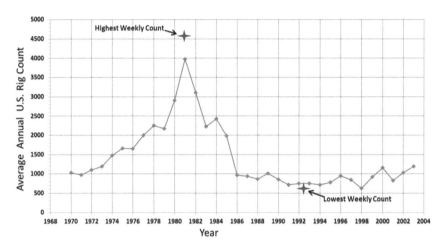

FIGURE 1.20
Steady rise and then fast fall of the oilfield with prolonged period before any major recovery.
(Baker Hughes Rig Count public data.)

4530 the last week of 1981 and started declining at an ever-increasing rate since the global economy was struggling toward a recession (Figure 1.20, shown as annualized data). The oil shortage and then the drag on most world economies would have significant effects on the oilfield and in the United States, which served to increase the need for more economic, but still effective, hydraulic fracturing stimulation methods. The lowest weekly rig count during this time was more than 10 years from the peak, with 596 rigs working in the week reported June 6, 1992. True recovery was not to occur until the Shale Revolution, which will be discussed much later, accounting for the rise we see here beginning to start after 1998, with the final week of that year reporting only 621 rigs working.

1.3.9.2 New Fracture Simulation Software Introduced

During the early 1980s, the growth of computing technology began to proliferate more rapidly. This provided an opportunity for those most interested in predictive modeling of hydraulic fracturing to have more intricate modeling and began to incorporate more formation data and data describing the fluid viscosity and leak-off characteristics, and for fracture simulation programs to run much faster. Additionally, for well operator and pumping service research personnel working on this need, government funding of university research also increased. By 1980, there were numerous early-stage fracture simulation models in the oilfield. With the Nolte-Smith analysis methods being used, better computer-aided analysis was also a strong driver for bringing computers into a more active part of onsite planning. By the mid-late 1980s, we began to see more digital data gathering onsite during stimulation treatments, bringing the need for incorporation of field data into our fracture models.

1.3.9.3 Horizontal Wells First Become More Common

Predicting production from a horizontal well is definitely more complex than for a vertical well because of the magnitude of the vertical permeability and the role it plays in production. In the case of vertical wells, the well is usually drilled to penetrate most, if not all, of the productive zone heights. Obviously, the presence of an aquifer or gas cap requires a special design. In the average vertical well, fluid usually moves horizontally or parallel to bedding planes; in other words, fluid moves radially toward the vertical well. In this case, the (often lower) vertical permeability would not have any effect on fluid flow unless the well only partially penetrated the formation. In the case of a horizontal well, fluid flow has to at least eventually move vertically toward the wellbore for the horizontal well to become productive.

Generally speaking, the vertical permeability of a formation is usually smaller than the horizontal permeability. This is usually because of the following three factors:

- Nature of formation deposition.
- Change in the type of deposited material. Usually, one would encounter thin layers of low-permeability fine sand between layers of higher-permeability coarser sand. The average permeability of the formation would be calculated in series, and the low-permeability layers would have a huge effect on the calculated average.
- Vertical stress is usually the highest stress.

Failure to acknowledge the effect of vertical permeability on the productivity of a horizontal well may lead to serious miscalculation.

1.3.9.4 Digital Electronics Become Dominant in Stimulation Oilfield Equipment

Digital pressure/rate data opened a door for greater understanding, storage, and use of data in simulation matching. Before the crash of the oilfield that saw drilling begin a free-fall event by 1984, the oilfield had experienced several years of "putting out fires" with respect to the methods often used. Fracturing technology growth was trying to take off but was actually hampered by being too busy and by the limited number of wells where digital data for observing formation response were available, which was only on a small percentage of well stimulations. However, post-1983 rigs were being stacked, while cementing and fracturing rolling equipment was being mothballed at staggering rates, and layoffs were high. Pumping service companies had no need to buy or build new rolling equipment for multiple years. Fortunately, we saw some pumping service companies take this opportunity to "digitize the oilfield" during this lull, and before 1990, we suddenly saw a new kind of oilfield service equipment that began to use digitized control and digitized data collection.

1.3.9.5 Late 1980s Bring the Digital Frac Van

With digital monitoring of the entire frac spread of equipment and treatment data, we began to see larger frac vans (Figure 1.21) as part of the standard expectation on the larger and more costly fracture stimulation treatments. In some cases, this van would also allow for a small fluid mixing lab onsite. This greatly enhanced the ability of well operators and fracture engineers to monitor and control the treatment.

1.3.9.6 Waterfracs Become Popular Again Because of Economics

Through the history of the oilfield up until the early to mid-1980s, vertical wells were about our only onshore completion type, especially with respect to fracture-stimulated wells. By the mid-1980s, we had observed a proliferation of massive hydraulic fracturing applications in many gas fields throughout the United States since gas prices were high and operators were trying to maximize production. Most of these fracturing treatments were using massive volumes of gelled fluids (even foamed fluids) to place treatments where a half-million pounds became a small operation. In some fields, it looked like a competition to be the operator who set the next record, with somewhere beyond 12 million pounds pumped into a single large interval occurring multiple times. This last record also coincided with the approach of the recession of 1984–85. Now we entered a time where lower gas prices began to show up and many of the MHF jobs in some applications were just not yielding adequate economic returns. What several of these gas fields began to experience was a "rebirth" of high-rate waterfracs using nongelled

FIGURE 1.21
Advent of larger monitoring vans were a major addition to applying fracturing technology onsite. (From Halliburton.)

or lightly gelled water and placing limited quantities of proppant in some of the gas reservoirs where permeabilities were very low, approximately 0.001–0.05 millidarcy (md).

1.3.10 Very End of 1980s and into Early 1990s

1.3.10.1 Coalbed Methane Rose in Importance Because of a Special Federal Incentive Program

As mentioned previously, fracturing coal seams to speed up degasification for safe mining was the earliest use of fracturing coal seams, but it was the 1990s before commercializing gas was being incorporated by mine owners. However, in the mid- to late 1980s, CBM wells completed in coal seams too deep for commercial mining became an important factor. Much of the sudden stimulus for this activity came from a vision to take advantage of a tax credit, generally called Section 29, which was part of the Crude Oil Windfall Profits Tax Act of 1980 to incentivize development of unconventional sources of gas or oil production, with methane from coal seams definitely falling under that definition in the early 1980s. However, in the early 1980s, there was an insufficient base of technology. In the later 1980s, we began to see coal from some of the smaller operators begin to successfully apply hydraulic fracturing stimulation successfully and focus on taking advantage of the tax credits. Originally, Section 29 tax credits could only be applied to wells drilled from 1979–1993 and to production only through 2002, and the magnitude of the credit was linked to gas prices of the production day.

With respect to one of the largest CBM fields to emerge, there was a sweet-spot area in the San Juan Basin where coal seams were thicker and had higher pore pressures and greater natural fracturing (cleats) that allowed them to be completed using an open-hole "cavitation" method, while other wells would require casing, perforating, and using hydraulic fracturing as the stimulation method. The tax incentive had been in place for several years before any large quantity of CBM wells were completed in nonmining application, with only two major fields representing the bulk of this activity: the Fruitland coals in the San Juan Basin of northwest New Mexico and southwest Colorado and numerous fields in the Black Warrior basin, principally in northwest Alabama. As there were very few large CBM fields to ever find large areas of high-pressure coals like the Fairway zone in the San Juan Basin, hydraulic fracturing became a staple of CBM wells. To illustrate the timeline for the rapid jump in CBM well activity after 1987, refer to the well count changed in Fruitland Coal in Figure 1.22. The Fairway open-hole wells were sometimes under-reamed or cavitated or both, while the cased wells were fracture-stimulated. Figure 1.22 also shows a map of the San Juan Basin CBM wells as of 1991 with the Fairway

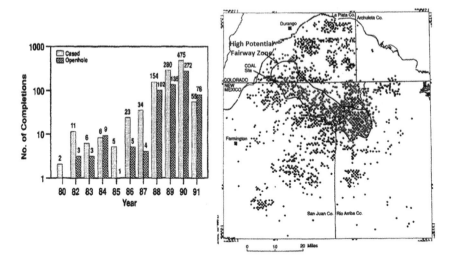

FIGURE 1.22
(Left) Number of CBM well completions in San Juan Basin through 1991. (Right) Map showing CBM wells with the Fairway trend identified. (Wikipedia download.)

enclosed where most wells could be "cavity completed" and produced at manyfold higher gas rates.

Even with the late start, there were more than 5000 CBM completions drilled and completed before the qualifying period for the tax credits expired.

Although CBM stimulation activity was active in the San Juan basin, the applications were mostly low technology (Diamond 1987; Jeu et al. 1988). However, in the Black Warrior basin, which became active for CBM wells later than the San Juan basin, there was a much greater variation and therefore more need for more technology in well completion and stimulation applications (Palmer et al. 1989; Palmer and Sparks 1989; McDaniel 1990).

1.3.10.2 More on Waterfracs

The CBM stimulation activity had brought a new interest in the use of waterfrac treatments with low sand concentrations. Additionally, when large amounts of gelled fluid were used in the fracs, some of these applications were in reservoir conditions where fracture cleanup was poor even with high sand quantities placed, and smaller volumes of sand placed with a waterfrac could actually produce better (Ely et al. 1988). The economics of waterfracs and lesser concerns for fracture cleanup continued to keep this stimulation process very active in some areas to almost inactive in others. When gas prices were good, it was difficult for many operators in lower-perm formations to do anything but "frac big." We will be discussing waterfracs again during the next cycle of lower prices.

1.3.10.3 Horizontal Wells Become More Widespread but Not Commonly Fracture-Stimulated ... Yet

During the late 1980s and into the 1990s, many were promoting horizontal wells in low-permeability formations as a substitute for needing costly fracture stimulation treatments. In general, this was successful in some applications but rarely in low-permeability gas wells where perms were sub-0.1 md or drilling costs were higher. In oil zones, it also faltered with perms less than approximately 2 md. Additionally, drill bits, steering, and other applicable tool technologies evolved slowly from 1980 to the mid-1990s with the limited number of horizontal wells drilled overall.

1.3.10.4 Oil Industry Gets a Huge Black Eye

Although it had no direct relationship with hydraulic fracturing, any catastrophic event in the oil patch has some accompanying negativity on the entire industry. On March 24, 1989, the Exxon Valdez oil tanker ran aground on Bligh Reef in Alaska, releasing nearly 11 million gallons of crude oil into Prince William Sound. It was the worst oil spill in US history at the time, but there were no deaths. Exxon reportedly spent more than $3.8 billion to clean up the site, compensate the 11,000 residents, and pay fines. Originally, the Alaskan court ordered the oil company to pay USD 5 billion in punitive damages in 1994. After 14 years of lawsuits and appeals, the US Supreme Court ruled that Exxon only owed $507.5 million in additional punitive damages.

1.3.10.5 Revisit of the Multiwell Experiment Site for the Gas Research Institute/Department of Energy-Funded Slant Hole Coring Project

The results of the fracturing applications to well MWX1 below 7000 ft were disappointing from a postfracturing production standpoint. However, it proved to the industry that we were not yet close to understanding what was happening with stimulation of tight gas sands where natural fracturing was extensive (Branagan et al. 1985). Additionally, the microseismic monitoring (MSM) data had borehole issues during the second fracturing treatment, so far-field data were not available for that treatment. However, there was confidence in the fracture directions for both zones fractured. The wellbore was at several degrees dip, so in the overhead view, they were in offset but parallel planes. Cleanup of the fracturing gel had proved to be an issue for the treatments, and this generated extensive fluid loss testing on cores to attempt to model the problems of gel fluid lost into natural fractures and helped improve reservoir production modeling in this type of reservoir.

In 1991, the original GRI MWX project site was revisited in the GRI/DOE Slant Hole Coring Project. Scores (probably >100) of technical papers present the massive volume of total data generated within the MWX Project, which began in 1983–1987, and include the revisit to the project site for the SHCP from 1991 to 1994. This is in addition to the complete project reports to the GRI/DOE, and many related technical papers continued through most of the 1990s. Even now, we still see scores of papers every year that reference one or more of those early papers (or the GRI/DOE project reports) and what was learned at M-site to support concepts or conclusions that may be presented for the first time even currently—and this trend will continue!

Actual observations were quite unlike the original expectations of many who want to have ultimate confidence in fracture simulation modeling. That is, once a single fracture plane was formed from the wellbore, it would then reopen to accept all later injections; there were multiple identifiable hydraulic fracture planes far away from the wellbore. It appears that with a formation with extensive natural fractures, the fluid can migrate and open these fractures when parallel to the hydraulic fracture path.

After retrieving and studying the core shown in Figure 1.23 from a sub-7000-ft deep zone in well MWX-1, the strange results observed in the fracture stimulations were now better understood. The degree of natural fracturing in the formation had been significantly underestimated (Warpinski et al. 1991).

Now, it was planned to accomplish similar work in this new project on the MWX-2 well in the upper A, B, and C sands between 4000 and 5000 ft, with more modern instrumenting of the other two wellbores. An added fourth wellbore would contain annular cemented-in geophones and also serve as the host well for the kick-off needed to perform the future slant hole coring in the B- and C-sands (Figure 1.24). There would be a postfracturing inclined

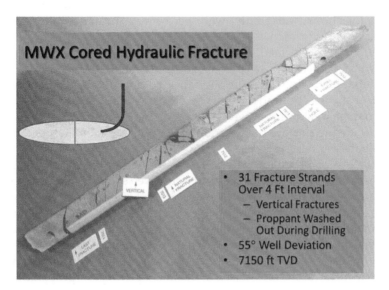

FIGURE 1.23
Core taken through the 1984 hydraulic fracture of the second treatment on well MWX-1. (From Pinnacle, Halliburton.)

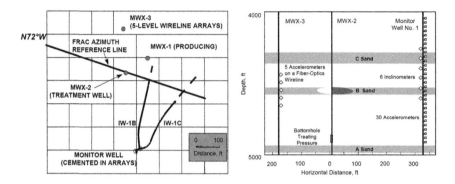

FIGURE 1.24
(Left) The SHCP revisit plan for the MWX-2 well at M-site; (right) B-sand fracture conditions. (GRI public data.)

core-through of the fractures in the B-sand and a prefracture inclined core hole (through the expected fracture plane) in the C-sand. The C-sand core hole would be examined, logged, and instrumented during all injections, and postfracture logging was also planned (Figure 1.24).

There were numerous microfracture tests performed (very small-volume, low-rate fracturing injections) to accurately determine the least principle stress at multiple locations from below the A-sand to above the C-sand. This was followed by a few injection tests in the A-sand performed before the other two wellbores were fully instrumented. The B-sand was the first zone for elaborate testing after all instrumentation was installed, and multiple fracture injections (water only) were performed and observed before the minifrac was performed and analyzed the day before the main fracture stimulation was performed. That would be the first time in this testing series that any proppant (sand) was placed (Warpinski et al. 1999).

The initial plan was to core through the hydraulic fractures in the B-sand and before any injections in the C-sand, with each coring to be approximately 120 ft from the wellbore; however, after seeing the core through of the B-sand, it was decided to precore the C-sand further out, approximately 200 ft from the wellbore.

One of the most informative physical results was when the fracture area of the B-sand was cored and examined. There were seven separate fracturing pump-ins over a multiday period, with all except the first initial breakdown expected to reach out to the area to be cored, approximately 121 ft from the wellbore.

The final three injections were the largest, with an approximately 17,000-gal. water-only injection test/falloff test followed by an identical linear gel injection-falloff test. Then, a day later, a 28,000-gal. main fracturing treatment was pumped, which was the only injection treatment placing any proppant sand.

During the B-sand core-through operations, which were expected to penetrate the fractured area approximately 120 ft from the wellbore, the coring barrel was run four times. The first two 20+-ft sections only exhibited a few cemented natural fractures, as expected; however, with approximately 18 ft of core run #3 completed, the core barrel was pulled early since it appeared that the cored rock had "twisted off." This event appeared to have occurred at the location of the propped fracture. There were a total of nine hydraulic fractures located within the core above the twist-off location, and none of the top nine fractures contained any sand. There had been significant proppant sand grains isolated from the coring fluid, further confirming the twist-off area had to be the propped fracture. The borehole inspection log run later confirmed two close but separate fractures at the core twist-off location, for 11 total fractures. Following wellbore washout, a fourth core barrel was run, but like the first two runs, found only a few natural fractures evident (they were slightly off vertical and had a different appearance when broken open). It seemed almost that most injection events had followed a separate path, or some followed multiple paths with fractures active long enough to reach out more than 120 ft to be found in the core-through (Figure 1.25). This was far less than the 31 identified fractures (with fluid residue) that had been identified in the core of the MWX-1 well below 7000 ft.

Before any injection into the C-sand, the prefracture core-through planned for where the hydraulic fracture should pass was redesigned to be further from the wellbore, at about 200 ft away instead of 120 ft like the core-through of the B-sand, in an attempt to be intersected by fewer fractures during the planned fracture injections and then the fracture stimulation treatment. The

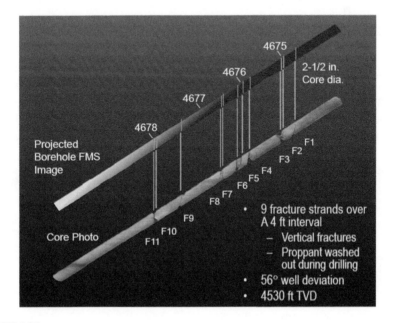

FIGURE 1.25
A 4-ft interval of the core-through of fractures found in the B-sand and the borehole image created from wellbore image logging. (GRI public data.)

cored area was logged with both sonic and imaging tools. The specific area where fracture penetration was expected was instrumented and isolated to a limited tunnel to reduce its volume and increase instrument sensitivity. In C-sand injections, the fracture approach, tunnel penetration, and fracture pressures were monitored before, during, and after fracture pass-through of fluid-only fractures and also during the propped fracturing treatment pumped. The fracture-intersected area of the core-through was also investigated postfracturing and again imaged to locate and evaluate fractures that had penetrated through the cored cavity.

1.3.11 Mid- to Late 1990s

1.3.11.1 Satellite Live Data Transmission from the Wellsite to the Electronic Host Center Comes to the Oilfield

The oilfield's high-pressure pumping service industry could see that it was not practical to introduce the high-performance computers of that era into its many hundreds of field-monitoring vans. Costs would be too high and few computers of that day could survive the trauma of many of the oilfield roads; however, the industry greatly desired to involve higher technology in real time on fracture stimulation of their wells. The short-term answer was to mobilize satellite dish transmission back to computing centers in cities. Figure 1.26 illustrates that answer.

1.3.11.2 Passive Microseismic Monitoring Becomes a Commercially Established, Generally Accepted Technology

During this decade, microseismic monitoring started allowing the industry to have a far-field visualization of fracture growth in vertical wells (Walker et al. 1998), but it was not yet considered a necessary tool for the limited market that included horizontal well stimulation. MSM had brought to light the importance of pressure sinks giving us mostly one-direction fracture versus

FIGURE 1.26
Satellite transmission of data to/from the city center analysis/monitoring site. (From Halliburton.)

our biwing beliefs when we would fracture near-depleted wells. Additionally, we would often see small fault zones take our fracturing fluid up or down to communicate with lower stress zones when we thought our wellbore zonal isolation was allowing us to choose where the fracture would grow. MSM was a great tool to help rebuild our faith in our basic rock mechanics and reservoir behavior beliefs and also to show us just how many natural flaws in these rocks there were that we did not expect. There were two major GRI-sponsored research projects that were both performed in the timeframe of 1997–1999: the Cotton Valley Hydraulic Fracture Imaging and Waterfrac Projects in east Texas and the Mounds Drill Cuttings Injection Project in northeast Oklahoma. Both of these projects would not only highlight the value of MSM in giving us data to show where fractures were growing, but also build confidence that onsite data interpretations with computational results within an hour or less could be trusted for decision making vs. the days to weeks an operator had been waiting to get data of any confidence.

1.3.11.3 Another Revival for Waterfracs

The early popularity of this low-cost, low-technology fracturing approach was aided by artificial economics of the industry's fracturing service pricing structures as they generally existed at this point in our history. In the latter part of the 1990s, we again see lower gas prices causing operators to start to back down from the massive fracs being placed in some of the low perm sand formations (Mayerhofer et al. 1997). It appeared clear that high-pressure pumping companies had been relying on chemical and proppant sales for nearly all their profit, as the pumping costs were often no more than break-even, and many multiyear service contracts had locked-in pricing structures. So with purchases of only a small amount of chemical and 15%–20% as much sand, this led to little or no service company profit margin when using higher horsepower and/or extended pump times. Suddenly, well stimulation costs could be greatly reduced by minimizing the chemicals and sand, and for a period of time, this allowed lower production rates to be much more economic. It was one to three years before these contracts expired and the pumping service pricing structures were altered to allow adequate profit on horsepower charges.

The economic advantage of waterfracs with low proppant volumes was less attractive once the pumping service contracts were all renewed, but by this time, low gas prices were forcing most operators to cut costs and live with lower production to keep completion costs under control (Mayerhofer and Meehan 1998). After a few years of data review, it became more clear that it was only the ultralow permeability zones, or zones where pore pressures were low or depleted, where the waterfrac approach proved to be the best economic answer. Many believe this is because these reservoirs are not able to adequately clean up following gelled fluid fracs, even though the propped fracture conductivity was higher, particularly if less damage to the fracture was achieved. The service industry began a period where better gel breaking, lower gel concentrations, and gels with fewer damage effects were the focus of products brought to the market.

1.3.11.4 Horizontals Increase, but without New Drilling Technology Since the Demand Was Low

As we approached a new millennium, onshore horizontal drilling applications expanded slowly among lower-permeability formations that might have been subject to fracture stimulation. This was likely due to the failure of the general industry to yet recognize the value of formulating the drilling plan to maximize the potential for fracture application within the lateral.

While horizontals were being drilled in many global locations, most of this growth was outside North America, and seldom were there plans to use hydraulic fracturing stimulation. When such a well did not show meaningfully better economics, it was a hard sell to US operators. On some occasions, the operators would then attempt to use horizontal completions with fracturing. Sometimes they saw great success, but moderate success was more common and uneconomic wells were not uncommon. The Bakken in eastern Montana and North Dakota was just such a hit-and-miss story up to this point in time and would be for several more years. Most operators at that time had not learned that consistent successful applications of fracturing a horizontal usually (or always, in many cases) required a surface-to-toe well plan focused predominately on *enhancing* the chances that fracture stimulation applications would be successful! Drilling costs were high and would stay that way until there was much higher demand, which was needed to interest the drillers in learning how better to drill and to offer newer technology to horizontal drilling. Additionally, coiled tubing use for stimulation of horizontals was still a quite limited event even in vertical wells, at least until the tail end of this decade.

Even in the late 1990s and into the early 2000s, very few operators believed that drilling the lateral in a specific direction was important; many operators would choose to drill parallel to the formation's preferred fracture plane (PFP) if fracture stimulation was planned, as this required less skill of the driller and crew and likely a lower cost to drill. Efforts specifically toward enhanced technologies for fracturing of horizontal wells were an area of application that was not yet being widely studied as a specific technology except by only a few within the stimulation service sector, and too few horizontals were being drilled to spur the drilling sector of the industry to extensive technology improvements.

A few major experimental horizontal fracture stimulation projects were completed, which did add to the science and technology but failed to make a compelling case to the investment community based on typical economic return. Even as this decade closed, only a small percentage of wells that were being drilled as horizontal completions were focused on a total drilling and

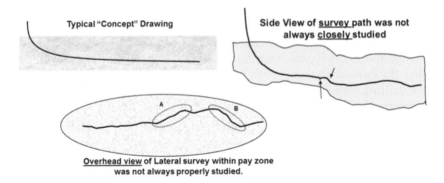

FIGURE 1.27
Concept drawing vs. realistic problems not recognized as important during the drilling of the lateral. (From McDaniel, B.W. 2007. A Review of Design Considerations for Fracture Stimulation of Highly Deviated Wellbores. *Paper SPE 111211 Presented at the SPE Eastern Regional Meeting,* Lexington, KY, October 17–19. doi:10.2118/111211-MS.)

completion plan dedicated to encouraging effective fracture stimulation, even though an increasing number of wells were drilled with anticipation of fracture stimulation as a plan, or at least a contingency plan. Too often, even the plotting of the wellbores used for review and decision making did not always use adequate density of survey data points to properly display the wellbore.

At this time, the primary reviews of wellbore path surveys were focused only on being able to run casing. Twists, turns, ledges, and dog legs were not often considered as having any effect on later fracture stimulation plans or were not considered as far as allowing room to run special tools. Even the perforating program could have been completed without concern for any effects that could have on fracture stimulation outcomes (McDaniel 2007). Figure 1.27 illustrates only a small representation of the scores of problematic cases encountered too late in the completion cycle and therefore having negative effects on overall fracture stimulation outcome.

With a backward view from our 2017 understanding of horizontal completion technology in low- to ultralow permeability formations, we would bluntly say that drilling horizontals in tight hard rock pre-2002 was simply clumsy in all respects. Operators did not yet understand what goals were most important; very few drilling crews knew how to best drill laterals; and they had only poor-quality drilling bits, muds, and steering tools.

None of this would change until the industry established a proven, large-volume need, and that is the next story in this chronology.

1.3.12 Late 1990 to 2002: How the Shale Revolution Started

Currently, we automatically link shale formation wells with drilling long horizontal laterals and massive multistage fracturing as the completion method; however, the first Barnett shale horizontal completion, drilled in by Mitchell Energy in 1992, was not economic, nor was it completed with a large multistage fracturing treatment. It was not drilled or completed in the fashion we think about 15+ years later. Before another shale horizontal well was drilled, the Barnett Shale in north Texas would have *already* become the most active drilling target in the United States ... as *vertical well* drilling targets! Yes, during the late 1990s, George Mitchell (Mitchell Energy Co.), after 15+ years of trying, was finally successful in making the Barnett Shale a highly economic vertical well gas play.

Some say this success in the Barnett shale came simply from using waterfracs (only water and friction reducer with low sand concentrations) instead of the various gelled fluids or foams. That is hardly the full story. Waterfracs had also been tried unsuccessfully multiple times in the Barnett years earlier, but success finally came when they decided to mimic the way waterfracs were most successful in the Cotton Valley Sands of east Texas, which was using massive volumes of water (up to a million gallons vs. approximately 100,000) and high injection rates, placing 100 mesh sand before switching to 20/40 mesh late in the stimulation treatment. Additionally, to ensure there would not be high water production, success was restricted to only drilling wells where there was adequate thickness of Viola limestone below the Barnett Shale to keep the fractures from communicating with the Ellenberger aquifer zone. This area, called the Core Area, was essentially the northeast part of the field and only accounted for approximately 40% or less of the potentially productive area of the Barnett Shale. Even so, this was a large area, and before the end of 2001, there were more rigs drilling in the Barnett Shale than any other field in the United States.

Another fundamental part of making the Barnett Shale a major gas field was the use of microseismic monitoring to allow Mitchell Energy (the major operator) to understand what was occurring during these massive waterfrac treatments (Fisher et al. 2002). Figure 1.28 shows the postfracture MSM map, and later it was found that five offset Barnett producing wells were drowned by this approximately million-gallon waterfrac treatment on this well. This term was soon called "bashing" and eventually altered well planning and the general stimulation approach.

We were not getting the simple biwing fractures we would generally observe in a fractured conventional formation in which we would pump these treatments. The Barnett Shale was very brittle and highly naturally fractured. Additionally, even at these depths of 7000–8000 ft, the difference in the least horizontal stress and the major horizontal stress was only 150–300 psi, allowing the natural fracture system to open enough to accept massive amounts of our fracturing water and create a huge area of communication for gas to flow. Without the understanding gained through MSM data, it is questionable that the field would have been such a fast-growing success. It seems very few fully understand the contribution made by MSM to the initial success of making the Barnett shale such a highly economic play. This

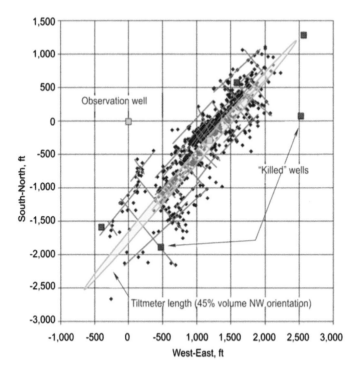

FIGURE 1.28

Barnett vertical well shale example where microseismic data shows complex fracture generation that resulted in five "killed" or "bashed" wells (red squares). (From Fisher, M.K. et al. 2002. Integrating Fracturing Mapping Technologies to Optimize Stimulations in the Barnett Shale. *Paper SPE-77441 Presented at the SPE Annual Technical Conference and Exhibition,* San Antonio, TX, 29 September–2 October. doi:10.2118/77411-MS.)

monitoring would later become equally important for learning how best to use horizontal completions (Figure 1.29).

After gaining a good understanding of this unusual response to fracturing, many of the Mitchell Energy engineers and geologists wanted to try horizontal wells, but George Mitchell would not approve it. After all, at that time, only 7% of the onshore rigs running in the United States were drilling horizontal wells. One major question they wanted to answer was, "Can horizontal completions allow us to drill economic wells outside the Core Area?"

1.3.12.1 Birth of the Horizontal Well Revolution

In 2001, Devon Energy approached George Mitchell with a USD 3.5 billion buyout offer of his company. At 82 years old, George decided to sell, and the deal was finalized in January of 2002. The "Father of the Shale Revolution" would retire and let others investigate new approaches to horizontal completions in

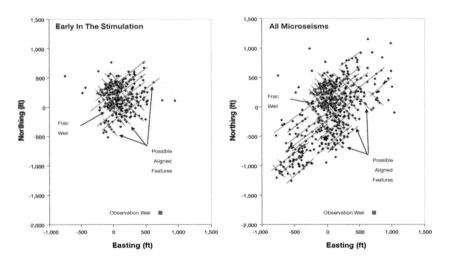

FIGURE 1.29
These two plots show the time-based development of aligned MSM data, indicating activation of natural fractures in a vertical Barnett Shale fracture stimulation. (From Fisher, M.K. et al. 2002. Integrating Fracturing Mapping technologies to Optimize Stimulations in the Barnett Shale. *Paper SPE-77441 Presented at the SPE Annual Technical Conference and Exhibition,* San Antonio, TX, 29 September–2 October. doi:10.2118/77411-MS.)

the Barnett. Following their purchase, Devon Energy (now using the same staff that had made the Barnett so successful) immediately began to make plans to evaluate horizontal completions in the Barnett. That summer, the first two horizontal Barnett Shale wells were completed, using a four-stage completion approach with approximately 800,000 gallon-sized high-rate waterfracs in each stage, and implementing a plug-and-perf approach that had become popular in many multizone vertical well completions. This was followed soon by two more similar horizontals completed that fall.

This method was not only a great success in the Core Area of the Barnett, but keeping the lateral relatively high in the Barnett zone, it was also successful outside the core, and now there was more than double the potential area for drilling economic wells in the Barnett. By 2005, the Barnett Shale field was drilling more than 80% of the wells as horizontal completions, with many other operators now drilling also. Originally, the major completion question was: Complete with cemented lateral or with open-hole packers to isolate open-hole sections? The technology of swell packers ultimately provided the premier solution for casing external isolation in open holes so that open-hole completions could be practical. Again, MSM is the technology that provides operators a real-time examination of completion effectiveness. Figure 1.30 shows MSM data evaluating stimulation effectiveness on a horizontal with swell packers isolating the noncemented wellbore into several sections for multistage completion, whether using plug and perf, a tapered ball-size ball drop, or a ball drop/sliding sleeve stage isolation method.

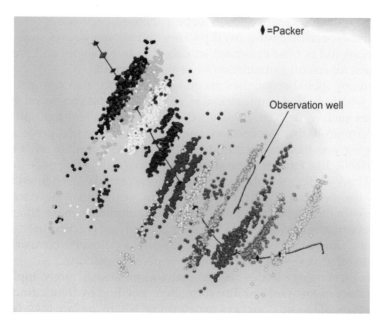

FIGURE 1.30
MSM monitoring multistage stimulation of horizontal with swell packers isolating the noncemented wellbore. (From Shaffner, J.T. et al. 2011. *The Advantage of Incorporating Microseismic Data into Fracture Models.* Society of Petroleum Engineers. doi:10.2118/148780-MS.)

By 2004, other US operators were beginning to examine other source rock shale reservoirs; at first, it was mostly gas, except for the Bakken in North Dakota. In the Bakken was one area that had drilled possibly the largest number of US horizontal completions in a single formation play, some in the eastern edge of Montana and some in North Dakota. However, it was not until 2005/2006 that the Bakken oil play fully adopted the shale completion approach to lateral placement, well design, and a multistage large volume/high pump rate fracturing stimulation model. Once this plan began to be followed, the Bakken play in North Dakota began a drilling boom never before seen in that area of the United States.

Other gas resource shale developments came on line quickly. We saw Fayetteville shale (2005), Haynesville shale (2007), Montney in western Canada, and Marcellus shale (2008) in Appalachia and the US Midwest. Once horizontal drilling became popular even outside the Barnett after about 2005, we began to see drilling technologies respond quickly with better tools, bits, muds, and so on, and rig crews became more experienced with both the formations and techniques of horizontal drilling. These things combined to quickly bring drilling costs down, further improving economics; however, we now saw the focus of well design being "how do we help improve the likelihood that we can successfully fracture-stimulate" since that was the one part of the operation that could not fail or the well would not be economic. The other physical tools were developed for both the well, perforating, cementing, plugs, and especially casing swell packers and sliding sleeves since many completions did not use cemented laterals. The technologies and pumping equipment for fracture stimulation were also improved.

With many global economies crashing in 2008, we also saw gas prices falling, which started a migration of most horizontal rigs to oil or rich gas zones such as Bakken, Eagle Ford, and the shales and tight oil sands in the Permian Basin. As expected, one thing that made that process more challenging was the tendency to drill and fracture-stimulate other source shales exactly like Barnett Shale wells. The stark truth is that none (to date) that still have sufficient hydrocarbons remaining are like the Barnett in two major ways: the Barnett is far more brittle than most and few have such low differences in their horizontal stress components. Even with those differences, operators still would eventually identify their own best practice through trial and error … and perseverance. Figure 1.31 shows a brittleness comparison of Barnett to other shales and a few other low-perm formations.

Another technology shales needed that was previously rarely applied by US operators in low-permeability formations needing hydraulic fracturing was to obtain 3D seismic surveys to determine where to place their laterals. This is even more useful if the well operator is able to construct a 3D earth model and know the ideal placement of the lateral section (Figure 1.32). We would get to log any zone in a vertical well before we complete it, but not so with horizontals. Even with drilling a pilot hole, with very long laterals, we need to know pretty close to where we want to place all of it before the bit turns horizontal for a mile or more.

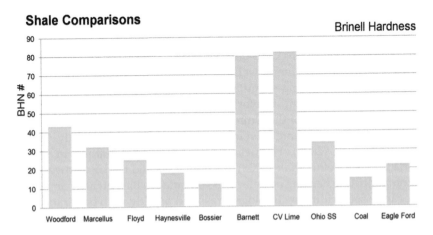

FIGURE 1.31
Brinell hardness comparisons, seven shales vs. a coal, Cotton Valley Lime, and Ohio Sandstone cores. (From Mullen, J. et al. 2010. Lessons Learned Developing the Eagle Ford Shale. *Paper SPE 138446 Presented at the Tight Gas Completions Conference in San Antonio, TX, November 2–3.* doi:10.2118/138446-MS.)

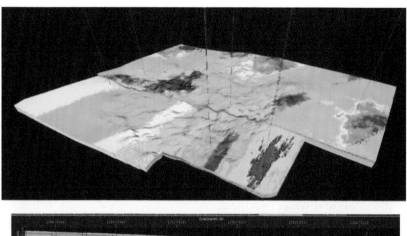

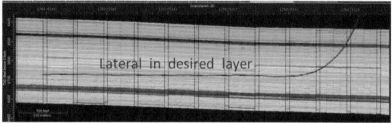

FIGURE 1.32
(Top) Large view of the reservoir, multiwell 3D view; (bottom) side view exhibiting lateral placement into the desired formation layer. (From Halliburton.)

The resource shale well timeline can be viewed as the following:

- Barnett Shale vertical wells prove to be commercial reservoirs by using large-volume, high-rate waterfrac treatments (mid-1990s).
- Microseismic monitoring helps to show that the highly complex nature of fracture systems was not a single hydraulic fracture as assumed, nor even a family of multiple, parallel hydraulic fractures (late 1990s).
- Extensive new core analyses declare Barnett gas-in-place 3 ft greater than earlier estimates (1997).
- George Mitchell is more than simply vindicated, he is proclaimed a wizard or genius—*the Father of the Shale Revolution!*
- In 2001, less than 7% of US rigs are drilling horizontal completions; only one (an uneconomic one in 1992) had been drilled in the Barnett Shale.
- Mitchell Energy is sold to Devon in late 2001.
- Sale of Mitchell Energy to Devon Energy completed in early 2001, and a new Barnett goal is established—evaluate *horizontal completions* to see if equivalent production of three to four verticals could be achieved at a cost of less than two vertical wells.

- Horizontals provided economic completions, even where the Viola lime was not present to protect the aquifer (~2003).
- Three-dimensional seismic surveys are determined to be an economic aid (a necessity in many cases) for candidate selection (previously were rarely used even among tight formations in the United States).
- By 2004, most Barnett Shale rigs are now drilling horizontals; many US operators start searching for the next equivalent to the Barnett Shale.
- Fayetteville shale (2005), Haynesville shale (2007), Montney (Canada), and Marcellus shale (2008) are developed.
- Increasing use of pad drilling proliferates to increase efficiencies, reduce footprint, and create an environmental impact of trucking materials and water handling.
- Bakken Shale converts to using the Shale Completion Model in oil resource rock; ignites leasing boom (2006–2008).
- Drilling efficiencies reduce completion times by more than half, even though laterals are being lengthened (2008).
- Questions such as, "Is waterfrac the best for non-Barnett shales?"
- Brittleness, a function of elastic modulus and Brinell Hardness of the rock, became a formation property of importance, as rarely have we found other US shales as brittle as Barnett (Figure 1.32).
- This difference in rock brittleness factor ultimately led to understanding that often, instead of using only waterfracs, *hybrid fracture fluid designs* (waterfrac early and gelled fluid late in stage) was the better choice (Figure 1.33).
- Horn River Basin in Canada and Eagle Ford in south Texas are both activated by application of the Shale Completion Model, and activity quickly becomes a drilling boom (2009).
- Permian Basin in West Texas begins to apply the Shale Completion Model to numerous low-permeability conventional oil formations with great success (2009–2010), launching several regional mini-booms.
- By early 2010, Barnett was the largest gas-producing field in the United States; more than 10,000 horizontal wells were completed in the previous eight years; approximately 55% of US rigs were drilling horizontal completions by the end of 2010.
- Numerous areas outside of North America began to apply the Shale Completion Model to low-permeability conventional and unconventional formations (2010–2011).
- Haynesville shale threatens to overtake Barnett as the number-one US gas-producing field; Bakken, Williston basin in North Dakota challenges to become the highest oil-producing area in the United States (end of 2011).

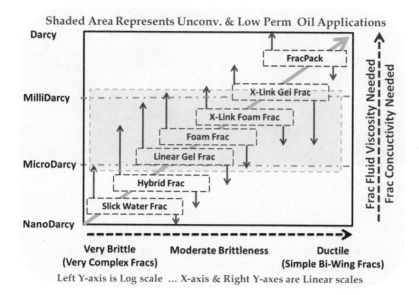

Left Y-axis is Log scale ... X-axis & Right Y-axes are Linear scales

FIGURE 1.33
Fracturing fluid selection defined by formation properties of permeability and brittleness. (From McDaniel, B.W. 2011. How "Fracture Conductivity is King" and "Waterfracs Work" Can Both Be Valid Statements in the Same Reservoir. *Paper CSUG/SPE 148781 Presented at the Canadian Unconventional Resources Conference,* Calgary, Alberta, Canada, November 15–17. doi:10.2118/148781-MS.)

- Eagle Ford oil and gas liquids plays draw many rigs from dry gas shales as gas prices fall (2011); more than 235 horizontal rigs are in Eagle Ford by mid-2012.
- In July 2012, 60% of US rigs are drilling horizontal completions; even 40% of the 42 rigs drilling in Barnett field are seeking oil as gas prices stay very low.

1.3.12.2 2010: Offshore Rig Fire and Spill of All Spills

In the Gulf of Mexico, with the seabed a mile below them, the greatest drilling disaster and oil spill yet to blacken any sea followed what happened on a quiet Tuesday evening April 20, 2010. Under contract to BP, the transocean rig *Deepwater Horizon* had completed drilling down to the Macondo zone 3 miles below them. The rig was preparing for cementing in the final string of casing, but there had been a growing group of events that weren't perfect, and they finally allowed the awesome strength of the oil-bearing formation to overpower the wellbore and then the surface safety equipment. Soon, fire engulfed the rig platform, killing 11 either in its path or from giving their lives trying to avert disaster; 17 others were injured, 4 critically, and approximately 100 escaped.

FIGURE 1.34
(Left) *Deepwater Horizon* rig after the fire had completely engulfed it. (Right) Rig beginning its dive to a water grave after 36 hours of the uncontrolled blowout burning. (Download from Wikipedia.)

After almost two days, the rig was still burning with the oil-fed fire still shooting upward when a great explosion occurred. Soon thereafter, the rig remnants finally sank, but oil was still rapidly spewing from the bottom of the sea to keep the fire alive a while longer (Figure 1.34).

Efforts to stop the flow now began. Having to work more than 5000 ft below water and after several methods were tried unsuccessfully, it was almost three months before the well would be effectively capped. By then, it had spewed an estimated 130 million gallons (July 15th).

As mentioned earlier concerning the 1989 Exxon Valdez tanker spill, there was no direct relationship to hydraulic fracturing involved, but *everything* connected to the oilfield was brought into question as to safety and environmental damage, particularly if related to drilling and completing wells.

1.3.12.3 Antifracture Activists Become Active over Potential Damage to Surface Water and Potable Underground Water Sands

Following the election of President Obama in 2008, the environmental activists had increased their voice and enlarged their following. One particular oilfield activity that had suddenly become an environmental target was "fracking," the media's term for hydraulic fracturing. The single most dramatic attack by environmentalists was in the form of a film written and produced by activist Josh Fox in 2010 only a short time before the Macondo Blowout event, which surely added public support for the film's "fake news" as is allowed in the film industry and also (improperly) bleeds over to much of the public media. The movie's theme was to attack oilfield activities and, most specifically, "fracking." The most dramatic movie scene shows igniting water flow from a kitchen sink and claiming this was the fault of a nearby gas well, although it was proven to be biogenic gas present naturally in the water well source sands years before the subject well was even drilled. This brought fame to the movie writer and promoted him to exalted status among environmental activists, especially those who supported the shutdown of the oil and gas industry in favor of completely natural energy sources such as wind and solar.

1.3.12.4 Hydraulic Fracturing Rarely Linked to Felt Seismic Tremors

Several states where oil and gas drilling is very active have experienced minor to moderate increases in earthquake activity starting about 2008. Anti-oilfield activists immediately stepped up to the soapbox to point fingers at hydraulic fracturing activities. There is some evidence that apparently links some of the increased seismic activity in a few oilfield areas to deep underground water disposal wells, especially in north-central Oklahoma and a few areas in Texas. Historically, such oilfield waters have been an everyday natural byproduct of oil or gas production. Several states have now began to more closely govern this deep water disposal more closely, occasionally stopping a few such disposal wells and reducing injection rates in others. However, the massive water volumes now being injected during hydraulic fracturing treatments are an easier and higher visibility target.

New research published in July 2017 suggests hydraulic fracturing and saltwater disposal has limited impact on seismic events. During a two-year study, University of Alberta geophysicist Mirko Van der Baan and his team (van der Baan et al. 2017) studied over 30–50 years of earthquake rates from the top six hydrocarbon-producing states in the United States and the top three provinces by output in Canada: North Dakota, Ohio, Oklahoma, Pennsylvania, Texas, West Virginia, Alberta, British Columbia, and Saskatchewan. With only one exception, the scientists determined no province or statewide correlation between increased hydrocarbon production and seismicity. This data review established that human-induced seismicity is less likely in areas that historically have fewer natural earthquakes. The anomaly was in Oklahoma, where seismicity rates have changed dramatically in the last five years, with strong correlation to saltwater disposal related to increased hydrocarbon production. The increased production itself is a direct result of increased hydraulic fracturing in the areas studied, but the actual hydraulic fracturing treatments have not been linked to increased earthquake activity.

1.3.12.5 Post-2010

Hydraulic fracturing and horizontal well drilling have given American companies access to vast new reserves of oil and gas and dramatically increased the production of hydrocarbons here in the United States. Since 2010, the United States has added approximately 5 Mbbl/D, and natural gas production is up approximately 33% during that same time period. The effects of this energy revolution have been felt globally—they've brought gasoline prices down for American drivers while remaking the global oil market. But here in the United States, they've been an enormous boon to an industry most Americans are likely unfamiliar with: petrochemicals. Cheap petrochemical feedstocks (a byproduct of oil and gas drilling) are pushing the US petrochemical industry to new heights.

Possibly the two greatest challenges to the upstream oilfield and use of hydraulic fracturing are *oil and gas market prices* as number one and possibly *environmental acceptance* as number two. The oilfield should not only continue to help improve the quality of its carbon footprint but should also strive earnestly to *enhance public acceptance* and not simply rely on simply regulatory acceptance.

1.3.12.6 Mexico, Argentina, China, and Australia Investigate Their Source Rock Shale Formations

Leading up to 2010, we saw global interest stirred in source rock shales. Some areas, such as several countries in Europe, have encountered successful populist movements to castigate the hydraulic fracturing process as an environmental problem, even though there is extremely little factual support for this at present. In countries where oil and gas developments are currently major industries, this has been only a minor issue at most.

In Mexico, where extensions of the south Texas Eagle Ford shale reach across the Rio Grande, the investigation of that resource was their first significant source shale experience. At the same time, the state oil company PEMEX also began to investigate the application of the horizontal shale completion method as an approach to helping improve production in the many low-perm oil fields within Mexico. Just as we have exhibited in areas within the Permian Basin in west Texas, this has proved to be successful, replacing numerous vertical wells with a single horizontal lateral with multistage fracturing during well completion of tight nonshale formations. Additionally, Mexico continues to investigate numerous shale source rock formations within its borders. While China has many potential shale source resources, the published efforts presently often give a challenging picture for some of the early target areas. Several are deep and complex target formations that will require large investments to evaluate economic potential.

1.3.12.7 Fracture Sand Becomes a Dominant Commodity and Is Often Handled as a Separate Well Service

With massive quantities of water and sand now major factors, each has developed into planning projects for horizontal well completions; however, contrary to water, the proppants are required to be transported from only a few origination points and are an item of commercial sale, shipping, storage, and final delivery to wellsite. By the turn of 2010, the increased oilfield demands for sand as a proppant had instituted annual sand supplier technical conferences, and smaller regional workshops dedicated to studying and reviewing the journey of this product to its ultimate oilfield usage were organized.

1.3.12.8 2015: Status of Hydraulic Fracturing

Although there is an abundance of gas and oil both in reserve and production, we believe that net production rates can be increased; however, much of our present oil and gas production and reserves exist in reservoirs that are difficult to explore and produce. Those reservoirs are combined together under the term "unconventional reservoirs." Unconventional reservoirs might be tight oil or gas, CBM, and (only in the recent two decades) resource shale oil or gas.

Tight oil and gas reservoirs that are defined as reservoirs with permeabilities in the microdarcy range would require fracturing stimulation to be economically viable. Application of massive hydraulic fracturing of vertical wells drilled in tight gas reservoirs was implemented in the 1970s and early 1980s, making those wells economic for gas prices of their day. Extensive drilling and multistage fracturing of horizontal wells in tight gas formations was not extensively applied and practiced until this century. Fluid flow of hydrocarbons in tight formations follows the standard Darcy equation as well as the non-Darcy flow equation for high-velocity flow inside a fracture.

CBM is a more complex system where gas exists as an adsorbed phase on the surface of organic carbon grains. Coals contain a system of cleats, with the degree of cleating dictating the productivity potential. Desorbed gas diffuses through the matrix of the CBM reaching the cleat system. One might use Fick's equation or other similar equations to represent the fluid flow inside the matrix of CBM. Fluid flow inside the cleat system, however, follows the Darcy equation. Coal cleats are usually a water-filled system, meaning that the cleat system may still be fully saturated with water. Pumping water out (dewatering the coal) is necessary before gas starts to desorb and flow into the cleats at adequate rates and then to the wellbore. It is highly recommended that diagnostic testing be conducted before dewatering the coal seam. In the case of CBM wells, any testing of the formation, whether it is well testing or diagnostic testing (MiniFrac/diagnostic fracture injection test [DFIT]) will have to happen shortly after the well is drilled and before fracture stimulation and the dewatering process start. Otherwise, the results of the analysis can yield cleat properties and may say nothing about the matrix itself. The properties of the matrix, such as sorption time, will have to be obtained using laboratory experiments. Tests performed after dewatering will not yield reliable results.

1.3.12.9 How We Achieve Economic Production from Shales

The numerous shale resource formations we are now drilling horizontally for oil and gas are naturally fractured and might behave as dual porosity/dual permeability systems. For both fracture stimulation designs and reservoir modeling, they might be simulated as a rock with discrete natural fractures or as dual porosity/dual permeability. Our resource shale formations will consist of ultra–low permeability matrix (in nanodarcies) and organic material with higher permeability natural fractures (in microdarcies). The organic material (TOC) may be as high as 10% of the total rock. The natural fractures and matrix would contain free hydrocarbons, while the organic material may have adsorbed gas on its grains. In this case, a shale formation may be considered a triple porosity model. Although the desorbed gas may contribute significantly to total production, the production profile does not usually have significant characteristics to help identify its contribution.

Because of the extreme reservoir conditions of resource shale formations, maximum reservoir contact is important to reach economic exploitation of the reservoir. Creating a large number of transverse hydraulic fractures from long horizontal wellbores has presently proven to be the best way to achieve economic success. This general well plan, often called the Shale Completion Model, combined with multiwell pad drill sites, has certain requirements to maximize economics (with some exceptions for exploratory drilling).

1.3.13 Components of the Shale Completion Model

- Multiwell pads sometimes might allow drilling one well while stimulating/completing other well(s) on the pad.
- Rig-free completion plans if possible (wells are often suspended after all tubulars are run, cementing is completed, and the wellhead is installed).
- Long lateral sections that are perpendicular to the maximum horizontal stress plane.
- Hold either within a few degrees of horizontal or following the formation dip when known.
- Completion plan is designed so as to maximize the potential for effective multistage hydraulic fracturing (number of stages and length of each completion stage).
- Determining the best fracturing stage design (pump rates, stage volumes, compatible fluids, proppants, proppant concentrations, etc.).
- Method(s) for fracturing stage isolation and post-treatment recovery plan for wellbore and fluids recovery.

1.3.13.1 *2017: Moderate to Low Global Oil Prices, Natural Gas Prices Low in the United States but Globally Higher*

Global oil price volatility is not simply supply and demand; it is also subject to supply issues that are not only political but also subject to war zones and safety of oilfield personnel. When it seemed the US oilfield needed more than USD 60/bbl to survive, after significant shrinkage of rigs and layoffs, oil shales have rebounded moderately with just a glimpse of USD 50/bbl oil, but when looking at USD 40/bbl, all are nervous about staying in business!

The liquefied natural gas (LNG) market appears to be the hope for low US natural gas prices, but years of lead time are required to deliver high quantities of LNG product to global markets.

> *The oilfield is not always an economically stable place, and next it could surge to new heights or be threatened to suffer badly again!*

2

Shale Gas and Oil Play Screening Criteria

2.1 Introduction

Among the organic-rich shale fields that underlie the United States and Canada, we chose 12 common successful shale plays, namely Barnett, Ohio, Antrim, New Albany, Lewis, Fayetteville, Haynesville, Eagle Ford, Marcellus, Woodford, Bakken, and Horn River. The shale gas and oil reservoirs currently being developed have been considered for decades as source rock. Multiple parameters, such as thickness, depth, TOC, thermal maturity (expressed as vitrinite reflectance), brittleness, mineral composition, total porosity, net thickness, adsorbed gas, and gas content are considered in the evaluation of shale plays, governing production from these assets. Curtis (2002) offers a geologically detailed description of most of the major shale plays. Le-Calvez et al. (2006) discusses how geological factors govern production from gas-bearing formations. Furthermore, Holditch (2006) introduces the notion of a resource triangle, whereby one should expect a lognormal distribution of resource quality in all oil and gas basins. Thus, in the major oil and gas provinces in the Middle East, there should be extremely large volumes of oil and gas in low-quality reservoirs. As such, any oil and gas basin in the world that has produced large volumes of oil and gas from conventional reservoirs throughout the last 100+ years should have oil and gas in even greater orders of magnitude in unconventional reservoirs yet to be discovered and developed (Holditch 2013).

2.2 Assessing Potential Reserves of Shale Plays

Geological, petrophysical, and geomechanical mapping are the best tools for screening the potential of shale plays. This screening strategy includes reservoir characterization, stimulation and completion strategies, and careful examination of the reasons for success in the development of other shale plays. Each shale has its own criteria and optimal method of development. It is important to observe that there is a direct relationship between fundamental properties of shale plays and their expected ultimate recoveries. These relationships can be used as a tool for evaluating shale play feasibility.

Tables 2.1 and 2.2 display the main and average characteristics of the major shale plays in North America.

Shale characteristics vary within the same basin. Completion conditions are similar to shale characteristics in terms of variations from one play to another. Completion designs for the 12 plays in the database include barefoot open-hole, nonisolated uncemented, preperforated liner, frack ports/ball-activated sleeves, and cased and cemented plug-and-perf. Lateral length and orientation, number and spacing of stages, proppant type (sand, resin-coated, ceramic, etc.), proppant mesh size and concentration, fluid type (slick water, cross-linked gel, hybrid, etc.), pump schedules/rates, and total volumes pumped per stage vary with operating procedures for different shale plays and also within the same play.

2.3 Shale Gas and Oil Production Criteria

Based on extensive experience with shale in North America, shale production characteristics vary within the same basin. The necessary elements for a shale gas play are as shown in Table 2.3 (Curtis 2002).

The necessary elements for a shale gas play are modified from Curtis (2002) as mentioned below. The main purpose of this table is to show the essential characteristics of any given shale for it to become economically profitable, with the ultimate goal of maximizing production.

2.4 Shale Evaluation Proposed Algorithm Data Structure

The main objective of this chapter is to evaluate new shale reservoirs relative to existing successful shale reservoirs based on the main geomechanical, petrophysical, and geochemical parameters. As a secondary goal, we aim to provide a database for the major productive shale plays in North America by building a metric that projects the various shale characteristics into one parameter (Euclidean distance). We envision that the proposed algorithm will function as follows. For each stored shale play's range of characteristics, the algorithm checks the input reservoir parameters and recommends the most appropriate shale to be used as an analog for future field development.

In order to accomplish the objectives, the chapter uses primary stages (components/steps) to build an algorithm such as including mineralogical,

TABLE 2.1

Major Shale Plays in North America and Their Main Characteristics

No.	Parameters Shale Plays	TOC Weight (%)	RO (%)	Total Porosity (%)	Net Thickness (ft.)	Adsorbed Gas (%)	Gas Content (scf/ton)	Depth (ft.)	Permeability (nanodarcies)	Geological Age
1	Antrim	5.50	0.50	9.00	95	70	70	1400	N/A	Upper Devonian
2	Bakken	10.00	0.90	5.00	100	59.50[a]	91.23[a]	10,000	N/A	Upper Devonian
3	Barnett	4.50	2.00	4.50	350	25	325	6500	25–450	Mississippian
4	Eagle Ford	4.50	1.50	9.70	250	35	150	11,500	100–2500	Upper Cretaceous
5	Fayetteville	6.75	3.00	5.00	110	60	140	4000	N/A	Mississippian
6	Haynesville	3.00	2.20	7.30	225	18	215	12,000	10–650	Upper Jurassic
7	Horn River	3.00	2.50	3.00	450	34	213.75[a]	8800	150–450	N/A
8	Lewis	0.45–1.59	1.74	4.25	250	72	29	4500	N/A	Devonian and Mississippian
9	Marcellus	3.25	1.25	4.50	350	50	80	6250	N/A	Devonian
10	New Albany	12.50	0.60	12.00	75	50	60	1250	N/A	Devonian and Mississippian
11	Ohio	2.35	0.85	4.70	65	50	80	3000	N/A	Devonian
12	Woodford	7.00	1.40	6.00	150	63.07[a]	250	8500	145–206	Late Devonian—Early Mississippian

Source: Modified from Curtis, J.B. 2002. *AAPG Bulletin*, 86(3), 1921–1938. Original data cited by the authors from the Gas Technology Institute.

[a] Entries were imputed using the regularized iterative PCA algorithm presented in Josse and Husson (2016) and implemented in the R package "missMDA" (see Section 2.4 for explanation).

TABLE 2.2

Average Shale Characteristics

Based on Analysis of 10,000 Shale Samples (Yaalon 1962)[a] (%)	
Clay minerals (mostly illite)	59
Quartz and chert	20
Feldspar	8
Carbonate	7
Iron oxides	3
Organic material	1
Others	2

[a] After Schön, J.H. 2011. *Physical Properties of Rocks: A Workbook: Handbook of Petroleum Exploration and Production*. 1st edition. Vol. 8. Elsevier.

TABLE 2.3

Essential Elements of Shale Gas for It to Be a Play

Laterally Extensive Shale	
Thickness	>100 ft. (Net shale thickness >75 ft.)
Total organic carbon content	>3%
Thermal maturity in gas window	(Ro = 1.1–1.4), Dry gas-Ro >1.0, wet gas-Ro = 0.5–1.0, oil-Ro < 0.5
Good gas content	>100 scf/ton
Moderate clay content	<40%
Brittle composition	Quartz and feldspar

mechanical, and petrophysical properties, along with the production indicators (e.g., gas content, adsorption, pressure gradient). These phases are the most important factors affecting the success of shale plays and feasibility of development. Consequently, the following are the primary phases of development used to build the algorithm:

1. Mineralogy comparison of shale gas and oil plays
2. Mechanical properties of shale gas and oil plays
3. Sweet spot identifier for shale plays
4. Production performance indicators
5. Sweet spot identification methodology (clustering model)
6. Spider plot of common shale plays' normalized petrophysical characteristics

2.4.1 Mineralogy Comparison of Shale Gas and Oil Plays

Mineralogical characteristics are the main sweet-spot proxy, used primarily in planning horizontal wells. The sweet spot is typically composed of higher

TABLE 2.4

Mineral Composition of Various Shale Plays in North America

No.	Shale Play	Quartz (%)	Feldspar (%)	Clay (%)	Pyrite (%)	Carbonate (%)	Kerogen (%)
1	Antrim	40–60	8.1[a]	10.1[a]	6.30[a]	0–5	8.89[a]
2	Bakken	40–90	15–25	10	6.20	5–40	12
3	Barnett	35–50	6–7	30	7	0–30	4
4	Eagle Ford		11–29	20	20	6.65[a]	4–11
5	Fayetteville	45–50	7.5[a]	15	6.597[a]	5–10	7.49[a]
6	Haynesville		23–35	20–39	29.5	7.12[a]	4–8
7	Horn River	9–60	0–3	53	7	0–9	7.13[a]
8	Lewis	56	3.6[a]	25	7.20	3.1[a]	8.78[a]
9	Marcellus	10–60	0–4	22.5	9	3–50	5.1
10	New Albany	28–47	2.1–5.1	17	6	0.5–2.5	7.71[a]
11	Ohio	42.7[a]	11.1[a]	36[a]	6.9	7–80	7.04[a]
12	Woodford	48–74	3–10	16	5	0–5	11.5

[a] Entries were imputed using the regularized iterative PCA algorithm presented in Josse and Husson (2016) and implemented in the R package "missMDA" (see Section 2.4 for explanation).

quartz and organic content (e.g., Alzahabi et al. 2015c). The brittle components are quartz, feldspar, and pyrite. However, the rest of the composition consists of ductile components. Table 2.4 shows the typical ranges for the mineral composition of various shale plays in North America.

2.4.2 Mechanical Properties of Shale Gas and Oil Plays

Table 2.5 shows general static mechanical properties in the vertical direction of the major shale plays. As is clear from the table, mechanical properties of these shale rocks vary significantly among the reservoirs and also within the same reservoir. The table does not reflect the variation in anisotropy that exists in most shale reservoirs.

2.4.3 Sweet Spot Identifier for Shale Plays

For each shale area, the sweet-spot segments of the resource are expected to contribute to successful treatment and development that could lead to higher recovery. These sweet spots control both the rate and volume of hydrocarbons (see Table 2.6). These sweet spot identifiers must be applied on a large reservoir scale and repeated to account for sweet-spot portions of shale rock and for well and fracture placement in shale rock.

2.4.4 Production Performance Indicators

The likelihood that some of the properties of a given shale play will ensure a reasonable success rate is based on the database, statistics, and a selection algorithm. Figure 2.1 shows a screening methodology by shale parameters

TABLE 2.5

Geomechanical Properties of Shale Plays

No.	Shale Play	Young's Modulus (million psi)	Poisson's Ratio ν
1	Antrim	3.107[a]	0.20[a]
2	Bakken	6	0.22
3	Barnett	3.5	0.20
4	Eagle Ford	2.5	0.23
5	Fayetteville	2.75	0.22
6	Haynesville	2	0.27
7	Horn River	3.64	0.23
8	Lewis	3.483[a]	0.21[a]
9	Marcellus	2	0.26
10	New Albany	2.903[a]	0.22[a]
11	Ohio	4.017[a]	0.23[a]
12	Woodford	5	0.18

[a] Entries were imputed using the regularized iterative PCA algorithm presented in Josse and Husson (2016) and implemented in the R package "missMDA" (see Section 2.4 for explanation).

TABLE 2.6

Average Sweet-Spot Criteria

Parameters	Conditions
Brittleness	>45% Rickman et al. (2008)
Young's modulus	+3.5 10^6 psi, Britt and Schoeffler (2009)
TOC	+2 weight %, Boyer et al. (2006)
Poisson's ratio	<0.2
Vitrine reflectance	Ro >1.3% for shale gas and <0.5% for shale oil
Kerogen type	Type I and II better gas yield than type III
Mineralogy	+40% quartz-calcite/less clay (less clay/low smectite <4 weight %)
Differential horizontal stress ratio (DHSR)	Very low <40%
Fracturability index	>65%, Alzahabi et al. (2015b)
Mineralogical index	>60%, Alzahabi et al. (2015c)

to judge maturity (Jarvie and Claxton 2002). The minimum cutoff used for differentiating mature versus immature shale play in Jarvie and Claxton's work is demonstrated for "A" and "B".

2.4.5 Sweet Spot Identification Methodology (Clustering Model)

The process first assesses maturity, then checks for similarity with major shale plays, then ranks and guides development strategy according to what has been learned from previous development of the existing 12 major shale

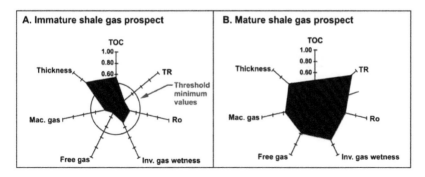

FIGURE 2.1
Normalized shale gas reservoirs of Utah. (Adapted from Jarvie, D.M., Claxton, B.L. 2002. Barnett Shale Oil and Gas as an Analog for Other Black Shales: Extended Abstract. *AAPG Midcontinent Meeting*, New Mexico.)

plays. A spider plot is used in this chapter to enable a view of all data points for comparison in one plot.

2.4.6 Spider Plot of Common Shale Plays' Normalized Petrophysical Characteristics

Figure 2.2 displays the characteristics of the 12 shale plays, normalized according to the highest values for each parameter. For example, Fayetteville is determined

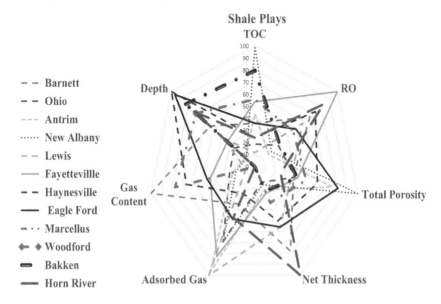

FIGURE 2.2
A spider diagram was constructed of the normalized shale resource characteristics. *Note:* Characteristics are grouped on the spider plot to illustrate the relationships that the algorithm identifies. Spider plot (e.g., Curtis 2002) is used as a tool to link different shale parameters.

TABLE 2.7

Ranges for Organic Content of Source Rocks

Total Organic Carbon (wt. %)	Kerogen Quality
<0.5	Very poor
0.5:1	Poor
1:2	Fair
2:4	Good
4:12	Very good
>12	Excellent

Source: Boyer, C. et al. 2006. *Oilfield Review*, 36–49.

to have the highest maturity among the shale plays, then other shale plays are normalized to its maturity value as a base reference. The characteristics of shale plays have an important role in determining production potential.

1. *Total organic carbon content:* It is a scale of organic content of source rocks, with the ranges shown in Table 2.7.
2. *Thermal maturity (%Ro):* Expressed as vitrinite reflectance, it is a measurement of the percentage of light reflected off the vitrinite maceral, where maceral is an organic component of coal. It is an indicator for differentiating among gas, condensate, and oil window.
3. *Total porosity:* The total porosity includes effective porosity and clay-bound water; where clay-bound water can be calculated from clay volume and clay porosity, then effective porosity can be easily estimated. The property is a significant factor in determining estimated volumes of oil and gas in shale plays.
4. *Net thickness:* Thickness is used as an input for calculating free and adsorbed original gas and original oil in place.
5. *Adsorbed gas:* The range in most shale plays is 20%–85% of the organic content of the shale.
6. *Gas content:* There is a direct relationship between gas content and TOC in shale.
7. *Depth:* According to Wang and Gale (2009), oil and gas operators have been produced in the United States from shale reservoirs 200–18,000 ft. deep.

2.5 Statistical Analysis of the 12 Shales

Standard techniques from multivariate statistical analysis were applied in order to attempt to quantify degrees of similarity in the 12 shale types. These techniques included multidimensional scaling (MDS), cluster analysis, and glyphs in the form of scatterplots and star plots. We will give an overview of the basic idea underlying these sophisticated methods, but the interested reader is urged to consult a reference on the subject, for example, Johnson and Wichern (2007), for a more detailed explanation.

In terms of quantitative promise, MDS is highest on the list since it relies on a measure of distance between cases (shales) in order to provide a tangible value of similarity. Since the data at hand are all quantitative, the usual Euclidean distance metric was used (in 15-dimensional space since 15 parameters were measured). MDS algorithms then seek a low-dimensional representation for the 15-dimensional data cloud while trying to preserve the relative distances. The clustering methods use the same metric to attempt to find similar groups of cases. However, different clustering algorithms will typically yield different results (clusters); thus, these methods are more visual in nature. The main types of clustering algorithms are: hierarchical, k-means, and model based. Finally, glyphs such as star plots provide only a visual comparison of cases by displaying the value of each parameter by means of a line emanating from the origin, the end result being a "star-like" or "spider-like" plot where cases with similar shapes would be deemed similar.

Thus, the figures in this section are various ways of displaying the similarities among shale plays. Their intent is to offer a quick visual assessment of their interrelationships; therefore, they should not be used as a rigorous measure of similarity. The only tangible metric that we propose in order to assess similarity between any given pair of shales is their actual Euclidean distance computed in the 15-dimensional space of corresponding parameter values.

2.5.1 Preliminary Data Preparation and Imputation

The data on the 12 shales had several missing parameter values corresponding to main characteristics (Table 2.1), mineral composition (Table 2.4), and geomechanical properties (Table 2.5). Imputation or "filling in" of these missing values is a necessity before attempting any substantive statistical analysis. Some of the best methods utilize regression-like approaches by which models are fitted to the entire data cloud in order to predict the missing entries. Among these models, the regularized iterative principal component–based algorithm discussed in Josse and Husson (2016) and implemented in the R package "missMDA" was chosen (R Core Team 2016). The results of the imputation for the 12 shales (plus two new shales, North African and Wolfcamp) are tagged with asterisks in Tables 2.1, 2.4, and 2.5. In all analyses below, shale parameters with a range of values were converted to single values by taking the midpoint of the range.

In the next section, the statistical analysis to produce a similarity analysis of the studied shale will be discussed. This similarity analysis will project the 15 dimensions of the studied shale in North America into one dimension by calculating the Euclidian distance. In order to proceed with MDS and clustering, and due to the different scales of measurement present for each of the 15 parameters, standardization was performed. For each parameter, this step involved subtracting the mean and dividing by the standard deviation calculated (empirically) from the 12 available values. This step has the added advantage of providing immediate meaningful comparisons of a particular parameter among different shales, since, for example, using the normal distribution as the frame of reference (standardized) values in excess of 2 or 3 units with different signs would suggest substantial differences between cases with respect to that parameter. For example, if shales 1 and 2 have (standardized) values of −2 and +3 for feldspar, respectively, then they have substantially different feldspar contents, since the former is 2 standard deviations below the mean (computed based on all 12 shales), while the latter is 3 standard deviations above the mean.

2.5.2 Statistical Similarity Analyses

Both k-means (partition of the points into k groups such that the sum of squares from points to the assigned cluster centers is minimized) and hierarchical clustering procedures suggested a natural grouping of the shales anywhere between two and four clusters. In k-means, a plot of number of clusters vs. within-groups sum of squares (SSE) reveals a rapid decrease occurring at these cluster numbers relative to 250 runs obtained by randomly permuting the data (Figure 2.3). This figure is used to determine the number of clusters (K). Strong clustering is indicated by an actual data SSE that decreases more rapidly than the 250 random runs, as cluster number increases, ideally with a knee-like feature observed at the optimum number of clusters (Figure 2.3). The absence of a pronounced "knee" means

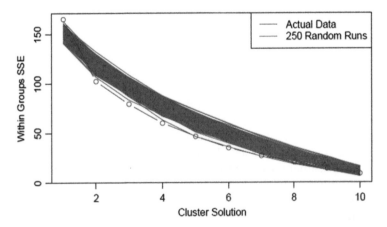

FIGURE 2.3
K-means cluster-validation plots.

that there is only weak evidence for clustering (for more details, see MacKay (2003)).

One could argue that there are two main clusters, with (Antrim, Bakken, New Albany, Woodford) forming the smaller group (Figure 2.4). The larger cluster is not so distinctly organized, but could ultimately be separated into three subclusters. In the ensuing analyses, we therefore color-coded the shale names according to this four-cluster grouping. The results of hierarchical clustering are displayed as a dendrogram in Figure 2.4. Distances between cluster centers are given by the nodes where two branches meet and can be read off the vertical axis. For example, the node connecting Antrim and New Albany is located at just below 3 units (2.88, in fact), which is the distance between them, while that for Bakken and Woodford is just below 4 units. Therefore, the node connecting the Antrim-New Albany and Bakken-Woodford subclusters (located at approximately 5 units) marks the distance between these two subcluster centers. Another visual rendering of the similarities is displayed in the star plot of Figure 2.5. The color-coding scheme in this, as well as subsequent plots, follows the four-cluster grouping discussed above.

A measure of the importance of each parameter in the separation of clusters can be obtained by running 15 different analyses of variances, one for each of the parameters, using the cluster grouping as the factor (with four levels since we are considering four separate clusters). The p-values (calculated probability) in Table 2.8 are arranged in increasing order and demonstrate that the

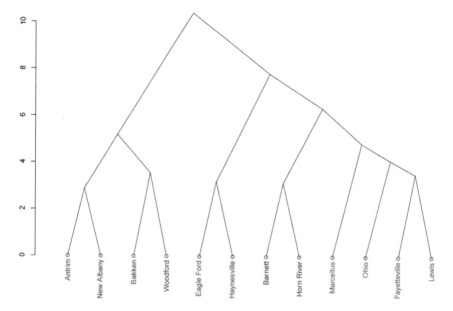

FIGURE 2.4
(Hierarchical) cluster dendrogram for the 12 shales.

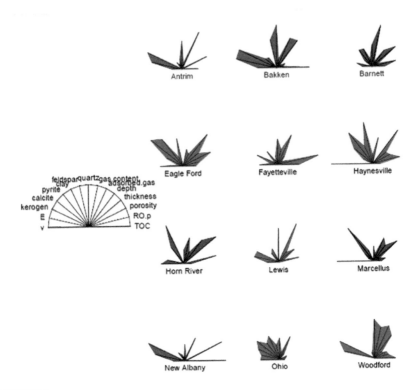

FIGURE 2.5
Star plot for the 12 shale plays.

TABLE 2.8

Sorting of the *p*-Values Determining Importance
of Parameters in Clustering Using the 14 Parameters

Parameters	Output *p*-Value for Each Variable (4-Cluster Solution)
Thickness	0.001048
Adsorbed gas	0.031291
Porosity	0.044572
Calcite	0.069496
TOC	0.083454
Depth	0.094142
Kerogen	0.094427
Clay	0.142634
Ro	0.157623
Quartz	0.174567
Pyrite	0.223213
V	0.361698
E	0.363506
Gas content	0.577719

TABLE 2.9

Sorting of the *p*-Values Determining Importance
of Parameters in Clustering

	Parameters	Output *p*-Value for Each Variable (4-Cluster Solution)
1	Kerogen	0.00016
2	Thickness	0.0011
3	Adsorbed gas	0.0118
4	E	0.0132
5	Feldspar	0.01339
6	Pyrite	0.0178
7	Quartz	0.02889
8	Porosity	0.0445
9	Calcite	0.0748
10	TOC	0.0834
11	Depth	0.0941
12	Ro	0.16
13	ν	0.16
14	Clay	0.18
15	Gas content	0.318

thickness is, being the most statistically significant parameter, contributes the most toward clustering, while gas content contributes the least (see Table 2.9). In other words, these results suggest that it is kerogen content that contributes the most toward the discrimination of shales into the four clusters depicted above (Figure 2.4). (Interestingly, for the two- and three-cluster solutions, kerogen remains the most important discriminating parameter.)

The cluster groups are displayed in yet a different way in Figure 2.6, the multidimensional scaling plot produced by the MDS analysis. Goodness-of-fit tests and stress measures indicated that, ideally, one should consider a 4-dimensional rendering of the 15-dimensional data cloud in order to minimize distortion in the intershale distances. However, the impracticality of this rendering, as well as the inherent difficulty in interpreting 3D visualizations, means that one is left with the usual 2D representation (Figure 2.6). (The stress values for dimensions 2, 3, and 4 were 8.0%, 3.4%, and 1.8%, respectively, corresponding to "fair," "good," and "excellent" quality renderings of the actual similarities between the shales in these respective dimensions; see Johnson and Wichern (2007), Section 12.6.)

The pairwise comparisons in Figure 2.6 constitute a more direct representation of the similarities. On the upper diagonal, we see the actual intershale distances from Figure 2.6, with font size inversely proportional to the magnitude (smaller distances appear in a larger font). The panels below the diagonal display scatterplots of the 15 pairs of parameter values.

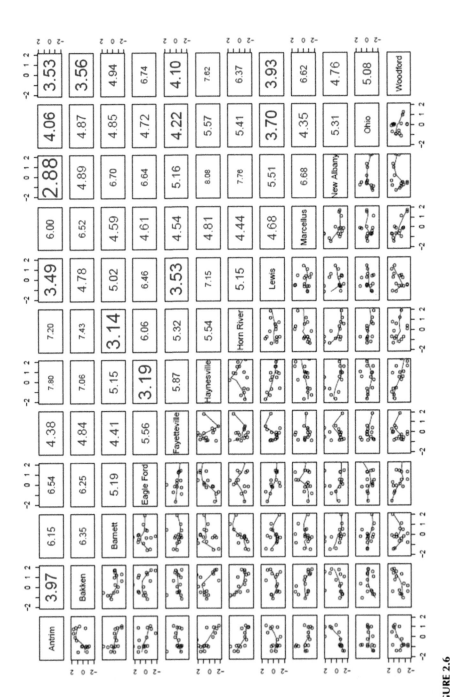

FIGURE 2.6
Parameter scatterplot and distances between the 12 shales.

Accordingly, the more similar the members of a pair of shales are in their values, the smaller their distance and the more closely the smoothed red line approaches a straight line through the origin with a slope of one (e.g., Antrim and New Albany). At the opposite end of the spectrum, very dissimilar shales will tend to have a scatterplot with a negative slope (e.g., Marcellus and Woodford). From this figure, one also sees the distortion in the MDS scaling plot of Figure 2.6, where Bakken appears closest to New Albany, but is in fact closest to Woodford.

2.5.3 Analysis of Two New Shale Types

In order to demonstrate the reliability and validity of the proposed algorithm, we considered two new shale reservoirs; the first is North African, while the second is a North American shale. Missing values for the 15 parameters corresponding to the new shales were imputed using the above-mentioned method of Josse and Husson (2016). The resulting values for the two new shales appear in Table 2.10.

This augmented 14-shale dataset was then subjected to the same MDS procedure used earlier for the 12 shales. However, because of the necessary standardization transformation that must precede MDS, we had two possible choices as to which mean and standard deviation to apply to the two new shales

TABLE 2.10

Values of the Tested Shale Plays

Shale	North African Shale	Wolfcamp Shale
TOC	2.3	2.3
Ro, %	1.3	0.96
Total porosity	10	2–10
Net thickness, ft.	300	1400
Adsorbed gas	37.9[a]	34.8[a]
Gas content	268.9[a]	324.9[a]
Depth, ft.	13,100	10,100
Quartz, %	45.3	61
Feldspar, %		
Clay, %	27.6	30
Pyrite, %	2.47	1.00
Carbonate (calcite), %	17.7	6.28
Kerogen, %	4.0[a]	5.0[a]
E, psia	6.5E + 6	4.6 E + 6
ν	0.25	0.24

[a] Entries were imputed using the regularized iterative PCA algorithm presented in Josse and Husson (2016) and implemented in the R package "missMDA".

within each parameter: (i) that for the 12 shales only, or (ii) that for the augmented 14 shales. Since the original 12 shales form the reference standard for comparisons with other types, option (i) is the most sensible. The implications of these findings will be discussed in the "Results and Flowchart" and "Conclusions" sections below. Table 2.11 gives the developed standardized Euclidean distances for the various shales, including the two new shale reservoirs.

Table 2.11 indicates that the Wolfcamp shale is not closely similar to any established shale reservoir. The North African shale has some resemblance to the Bakken and Haynesville shales.

2.6 Horizontal Completion Fracturing Techniques Using Data Analytics: Selection and Prediction

This section targets a comprehensive predictive model to evaluate the key success of completion strategies (treatment) for the major successful shale plays and guide future selective optimum completion for each shale play. Many important parameters that control producing well behaviors such as number of horizontal wells, spacing between fractures and wells, horizontal well completion configurations, stages per well, fracture type, average water requirement, depth, proppant type, hydraulic horsepower (HHP) per stage, Lb/ft^2 of proppants per stage, number of stages, and lateral length of the horizontal wells have been analyzed.

The proposed analysis is performed on the 12 major shale gas and oil plays for which the data were available. The analysis of the data identified similarity in completion strategies. Learning from these analyses can be used to predict completion strategies in new wells of old or new shale plays.

A case study from Niobrara shale (Colorado) is investigated. The procedure used in exploring the case study can be used as a decision criterion for similar cases in deciding stimulation configurations and main important factors that lead to the optimum way of developing these resources. Principal component analysis (PCA) is used to correlate the commonly used completion strategies with the geochemical and geomechanical properties of shale rocks.

2.6.1 Use of Big Data in Predicting Completion Strategies

Many horizontal wells and fracture stages are needed to drain a shale reservoir, and there is a need for effective fractures and horizontal wells to produce these reservoirs. We conducted a case study of the use of data analytics to study the Niobrara shale rock and suggest the best completion strategies used to effectively produce from the shale rock.

TABLE 2.11

Standardized Euclidean Distances between the 14 Shales

	Antrim	Bakken	Barnett	Eagle Ford	Fayetteville	Haynesville	Horn River	Lewis	Marcellus	New Albany	Ohio	Woodford	North African
Bakken	3.97	–	–	–	–	–	–	–	–	–	–	–	–
Barnett	6.15	6.35	–	–	–	–	–	–	–	–	–	–	–
Eagle Ford	6.54	6.251	5.18	–	–	–	–	–	–	–	–	–	–
Fayetteville	4.37	4.84	4.41	5.56	–	–	–	–	–	–	–	–	–
Haynesville	7.802	7.06	5.15	3.19	5.87	–	–	–	–	–	–	–	–
Horn River	7.19	7.43	3.14	6.06	5.32	5.55	–	–	–	–	–	–	–
Lewis	3.48	4.78	5.02	6.46	3.52	7.15	5.14	–	–	–	–	–	–
Marcellus	5.99	6.53	4.58	4.61	4.54	4.81	4.44	4.68	–	–	–	–	–
New Albany	2.88	4.88	6.69	6.64	5.16	8.08	7.75	5.52	6.67	–	–	–	–
Ohio	4.063	4.86	4.85	4.72	4.22	5.57	5.41	3.71	4.35	5.31	–	–	–
Woodford	3.53	3.56	4.94	6.74	4.09	7.62	6.37	3.93	6.62	4.76	5.08	–	–
North African	7.86	6.98	7.26	7.36	8.08	7.03	7.94	7.93	8.89	8.65	7.68	7.22	–
Wolfcamp	13.68	13.04	11.87	12.96	13.64	12.56	11.94	12.87	13.28	14.51	13.67	12.83	9.33

We then compared the characteristics of the Niobrara shale rock with the 12 most common shale plays in North America. Geochemical and geomechanical properties of these shale rocks vary significantly among the reservoirs and within the same reservoir. We believe that this chapter will assist in selecting proper completion techniques for horizontal wells.

It is a standard belief in the industry that the sweet-spot segment of the rock leads to successful treatment and then a higher recovery. Many authors have used data mining and data analytics to extract hidden relations and patterns between various parameters and sweet spot locations. Jansen and Kelkar (1996) use exploratory data analysis of production data to understand the relationships that govern fluid flow. Grieser et al. (2006) use data clustering techniques and self-organized maps to highlight completion and reservoir data that mainly affect production from the Barnet shale. This was archived through extracting useful information by reducing the statistical noise. Modeland et al. (2011) analyze data including number of frac stages, average treatment rate, total proppant amount (Sacks), average proppant per stage (Sacks), entire treatment proppant concentration (PPG), fluid type, and total number of clusters from Haynesville shale wells of eastern Texas and northern Louisiana. Their work shows that the increase in number of stages, conductivity, and proppant concentration directly increased the production. Cross-linked gel in this analysis has a slight advantage in 12-month production.

Slocombe et al. (2013) implement data analysis to group similarity stressed rock for treatment that led to the increase in the number of clusters that contribute to production of the Eagle Ford. An increase of 28% in the efficiency was obtained in the wells in comparison with their offsets. Chaudhary and Lee (2016) establish a method to detect outliers in rate and pressure data used for decline analysis and forecasting. Lehman et al. (2016) use big data in analyzing drilling data through the use of neural networks to optimize completion.

Based on these efforts and assuming no operational or human preference of certain completion strategies, a benchmark for all shale plays could be established. The benchmark may be linked to production trends or main reservoir parameters that affect the production from shale plays.

2.6.2 Data Analytics: Collection and Management

Reservoir data (geomechanical, petrophysical, and geochemical) collection is very important to project success in designing completion systems of horizontal wells. They work as an input for proxies used to place wells and fractures in reservoir models. These sweet spots typically consist of high Young's modulus, low Poisson's ratio, high kerogen content, high TOC wt. %, high fracturability indices, and lower clay content. Data points of 14 geomechanical, petrophysical, and geochemical parameters representing average characteristics for the 12 shale plays are listed in Tables 2.1 through 2.5.

A clear understanding of these data is important in achieving the objective of designing efficient completion systems, especially in shale reservoirs. Building a database of all successful cases of the current practice of completing horizontal wells could give us an idea about how to complete wells in a similar fashion to the successful completions in shale resources. Enough representative data would lead to good predictability models. The use of these models in estimating and predicting the completion systems of newly drilled wells is expected to achieve a high rate of success, especially in new and developing shale plays.

2.6.3 Statistical Analysis

The pairwise comparisons mentioned in Figure 2.6 and reproduced in Figure 2.7 constitute a more direct representation of the similarities, including the Niobrara shale. On the upper diagonal, we see the actual intershale distances, with font size inversely proportional to the magnitude (smaller distances appear in larger fonts). The panels below the diagonal display scatterplots of the 14 pairs of parameter values. Accordingly, the more similar the shales in a pair are in their values, the smaller their distance and the more closely the smoothed red line approaches a straight line through the origin with a slope of one (e.g., Bakken and Woodford, with a distance of 3.08). At the opposite end of the spectrum, very dissimilar shales will tend to have a scatterplot with a negative slope (e.g., Marcellus and Woodford, with a distance of 7.60, are the farthest apart among the 12 plays). Precise details of these distance calculations are given in Section 2.4 and Appendix D.

2.6.4 Analysis of Niobrara Shale Formation for Completion Strategies

To demonstrate the reliability and validity of the proposed algorithm for completion-style prediction, we considered one new shale reservoir, Niobrara. The Niobrara shale formation is located in northeastern Colorado with parts of the formation trending into Wyoming, Nebraska, and Kansas. Niobrara is an organic rich shale, varying from high carbon content on the east side of the play to higher clay content on the west side of the formation. The missing value for the Adsorbed Gas parameter was imputed using the above-mentioned method of Josse and Husson (2016). The resulting parameter values for the Niobrara shale play appear in Table 2.12.

This augmented 13-shale dataset was then subjected to the same MDS procedure used earlier for the 12 shales. However, because of the necessary standardization transformation that must precede MDS, we had two possible choices as to which mean and standard deviation to apply to the new shale within each parameter: (i) that for the 12 shales only, or (ii) that for the augmented 13 shales. Since the original 12 shales form the reference standard for comparisons with other types, option (i) is the most sensible. Table 2.13 gives the standardized Euclidean distances between the various shale reservoirs (see Appendix D for details of the computation).

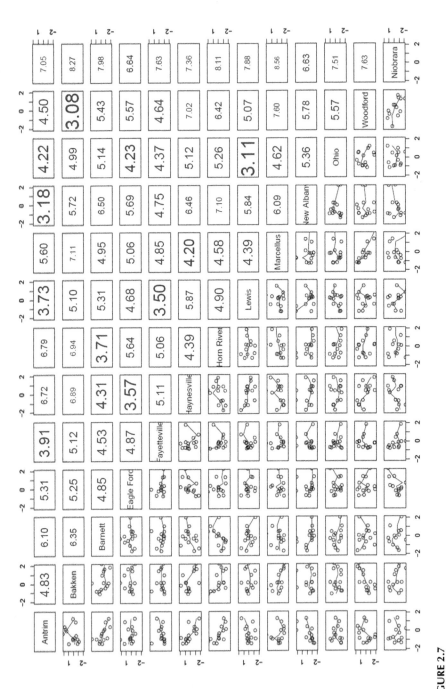

FIGURE 2.7
Parameter scatterplot and distances between the 12 shales in addition to the tested shale play (Niobrara).

TABLE 2.12

Values of the Tested Shale Plays

Shale	Niobrara Shale
Age	Cretaceous
TOC, wt. %	4.84
Thermal maturity, % Ro	0.6–1.30 (0.6)
Total porosity	9.35
Net thickness, ft.	275–400 (337.5)
Adsorbed gas	39.6059[a]
Gas content, Scf/ton	15–40 (27.5)
Depth, ft.	6800–7300 (7050)
Quartz, %	25
Feldspar, %	
Clay, %	15
Pyrite, %	1.5
Carbonate (calcite), %	46
Kerogen, %	6
Young's modulus, E, million psia	2.405
ν, Poisson's ratio	0.235

[a] Entries were imputed using the regularized iterative PCA algorithm presented in Josse and Husson (2016) and implemented in the R package "missMDA".

Table 2.13 shows (standardized) Euclidean distances between the 13 shales. The analyses indicate that the Niobrara shale is closest to New Albany ($d = 6.628$) and Eagle Ford ($d = 6.635$).

2.6.5 Results: Selection of the Completion Strategy

The model may work as a predictive tool to evaluate the key success of completion strategies (treatment) for major successful shale plays and guide future selective optimum completion for each shale play. Many important parameters control production from shale wells, such as number of horizontal wells, spacing between fractures and wells, horizontal well completion configurations, stages per well, fracture fluid type, average water requirement, depth, proppant type, hydraulic horsepower per stage, 1B/ft^2 of proppants per the stage, number of stages, and lateral length of horizontal wells. The proposed analysis performed on 12 shale gas and oil plays for which the data were available identified similarities in shale characteristics that may lead to similarity in completion strategies. These analyses can be used to predict completion strategies in new wells of old or new shale plays.

A case study from the Niobrara (Colorado) formation was investigated. The procedure used in exploring the case study can be used as a decision

TABLE 2.13

Standardized Euclidean Distances between the 13 Shales Including Niobrara

	Antrim	Bakken	Barnett	Eagle Ford	Fayetteville	Haynesville	Horn River	Lewis	Marcellus	New Albany	Ohio	Woodford
Bakken	4.830719	—										
Barnett	6.095365	6.352237	—									
Eagle Ford	5.307857	5.247100	4.847178	—								
Fayetteville	3.905588	5.123783	4.528308	4.868270	—							
Haynesville	6.715155	6.893276	4.308868	3.565157	5.108837	—						
Horn River	6.785320	6.936655	3.709059	5.641707	5.055577	4.394323	—					
Lewis	3.727838	5.097371	5.310933	4.681848	3.497240	5.874422	4.896284	—				
Marcellus	5.601065	7.112691	4.946537	5.063785	4.852775	4.203224	4.582889	4.394225	—			
New Albany	3.182359	5.719667	6.495983	5.686248	4.753613	6.459041	7.101061	5.843601	6.094538	—		
Ohio	4.221939	4.988114	5.141170	4.227655	4.367463	5.118648	5.262600	3.107880	4.624905	5.361957	—	
Woodford	4.504680	3.081079	5.425362	5.569794	4.640504	7.020888	6.423348	5.068525	7.595233	5.784913	5.565036	—
Niobrara	7.054853	8.272013	7.975972	6.635483	7.625429	7.363259	8.107038	7.879227	8.557301	6.628269	7.510023	7.628096

TABLE 2.14

Completion Strategy Recommendation Based on Euclidean Distances between the 13 Shales

Major Play	Shale Play	Average Rate	Primary Fracture Type	Average No. of Stages	Completion	Primary Proppant Mesh	Hydrocarbon Type
1	New Albany	40	Hybrid	12	PnP	30/50, 40/70, 20/40	N/A
	New Albany	35	WF	N/A	Fracture valves/PnP	30/50, 20/40	N/A
2 Yes	Niobrara	45	Hybrid	30	Fracture valves/PnP	30/50, 40/70, 20/40	Oil/Gas

criterion for similar cases in deciding stimulation configurations and the main important factors that lead to the optimum way of developing these resources. Principal component analysis is used to correlate the commonly used completion strategies with geochemical and geomechanical properties of shale rocks. The Niobrara shale characteristics (see Table 2.12) differ in several respects from those of the 12 major US shale plays; however, it can be concluded that the test shale is most similar to New Albany ($d = 6.628$) and Eagle Ford ($d = 6.635$) based on this parameter analysis. As an example, cased hole completion categories and fracturing methods are given for cased hole completion in Figure 2C.1 (see Appendix C).

Table 2.14 shows the Niobrara practice versus the New Albany completion practice. They look very similar in terms of the completion strategies of the horizontal wells.

Figure 2.8 shows a scatterplot of the two principal components (PCA) of the parameters Average Rate and Average No. of Stages (first PCA vs. second PCA) for the shales in Table 2C.2 (Appendix A), and can be used to reveal the presence of possible clustering; shales closer together are more similar (with respect to the two parameters) than shales farther apart. In fact, we see evidence of a large cluster centered approximately at a first PCA value of −1 and second PCA value of 0.

Figure 2.9 shows pairwise scatterplots (nine panels) of the three parameters Average Rate, Type/Fracture/Fluid, and Average No. of Stages for the shales on Table 2C.2—Appendix C. The three panels along the diagonal display distributional summaries appropriate to the type of the data composing each of the three parameters: numerical for Average Rate and Average No. of Stages (probability density plots) and categorical for Type/Fracture/Fluid (bar chart). The five-color coding scheme (one color for each value of Type/Fracture/Fluid) is consistent throughout; for example, red is used to represent shales with Type/Fracture/Fluid of Hybrid.

The leftmost bottom panel is a scatterplot of Average Rate vs. Average No. of Stages, with regression line fits through each of the five clouds of points.

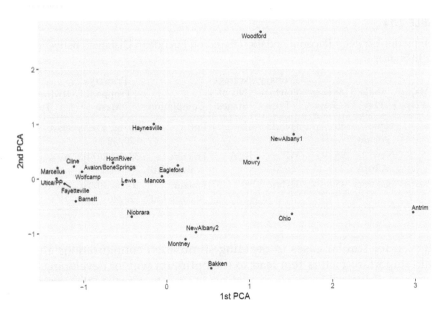

FIGURE 2.8
PCA plot of average rate and stages.

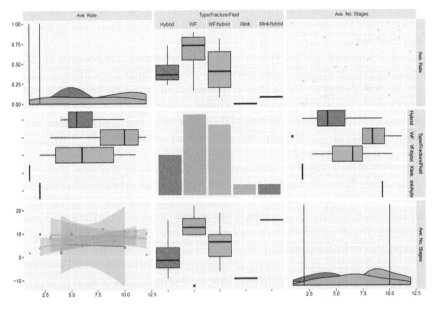

FIGURE 2.9
Pairs plot of average rate and number of stages.

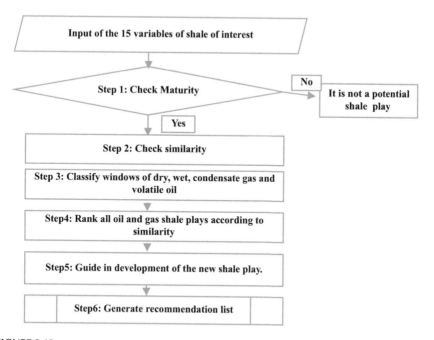

FIGURE 2.10

Shale algorithm flowchart. Step 2 uses Euclidian distance (Section 2.4), step 3 vitrine reflectance, and step 4 rank-based Euclidian distance.

The topmost middle panel displays by-color boxplots of Average Rate for the shales that fall into each of the five color groups.

A flowchart of implementation (Figure 2.10) of the proposed approach and recommendation on one of the horizontal wells from Niobrara is given in Appendix A. A comparison of average of production of eight horizontal wells from Niobrara is given in Table 2.13. It shows a 47% increase in production by implementing our recommended completion strategy.

2.7 Results and Flowchart

The flowchart given in Figure 2.10 was used in two case studies to assess the potential of each shale play: case study 1 is for the North African shale play; case study 2 is for the North American shale play (Wolfcamp shale play). Testing new shale characteristics requires several initial data points, including the main petrophysical properties (e.g., TOC, thickness of zone of interest, maturity). Appendix B offers a detailed comparison of the two shale reservoirs under investigation with the12 commercial shale reservoirs of North America.

2.8 Conclusions

This chapter demonstrates the proposed algorithm for screening existing and new shale plays. All in all, this study can be summarized as follows:

- All successful shale plays in North America have been identified and classified into clusters, with the performance of each compared according to 15 parameters characterizing the shale, followed by an assessment of their similarity using Euclidian distance.

- A guiding database for major productive shale plays in North America has been developed to list all potential development approaches, with guidelines suggested in order to identify the sweet spots in unconventional resources.

- A new shale evaluation methodology and candidate selection algorithm have been developed. The algorithm is based on the main geomechanical, geochemical, and petrophysical parameters of the newly discovered shale. It functions as a guideline for identifying sweet spots and identifies operationally approved methods from analogous reservoir development in order to increase the potential recovery of existing shale natural gas and oil accumulations.

- The statistical analysis suggests that kerogen contributes the most toward the grouping of shale plays into clusters.

- With the use of Euclidian distance based on a shale play's characteristics as the measure of similarity, Antrim and New Albany are the two most similar shales, whereas Antrim and Haynesville are the most dissimilar.

- Application of the proposed algorithm to the two new shale plays (Wolfcamp and North African) reveals that the North African is somewhat similar to Bakken and Haynesville in characteristics, whereas Wolfcamp is very dissimilar to all North American shale plays. Application to Niobrara reveals that thickness is the most important parameter in deciding on the completion strategy.

Appendices

Appendix A: Abbreviations

H	thickness
$\varnothing$	porosity

C	carbonate
Cl	clay
DHSR	differential horizontal stress ratio
K	permeability
LS	limestone
MDS	multidimensional scaling
Q	quartz
SSE	sum of squared errors of prediction
Ro	vitrinite reflectance (a scale of maturity)
TOC	total organic carbon
TR	transformation ratio or extent of organic matter conversion into hydrocarbons
Mac. gas	volume of macerated cutting headspace gas, a measure of gas yielded upon fracturing of Barnett shale
Inv. gas wetness	gas composition, such as gas wetness
PnP	perf and plug fracturing method, where pump-down bridge plugs and perforating guns are used to both isolate previously fractured intervals and perforate new intervals for the next hydraulic fracturing treatment

Appendix B: Analysis of Two Cases of Shale Plays

Case 1: North African Shale

As an example to test the system, a model for a shale reservoir comes from North Africa. The North African shale characteristics (see Table 2.10) differ in several respects from those of the 12 major US shale plays. Figure 2B.1 is a thickness clustering comparison with the 12 shale plays in the database. From the thickness indicated by the spider plot, it can be concluded that the test shale is very similar to the Barnett and Eagle Ford shale plays. Figures 2B.1 through 2B.6 show the comparison between characteristics of North African shale and those of the major shale plays. Figure 2B.2 demonstrates the TOC spider plot comparison, according to which Ohio is the closest in TOC similarity to the test shale.

Case 2 North American Permian Basin Wolfcamp Shale

Properties of the Permian Basin's Wolfcamp shale reservoir are listed in Table 2.10. Furthermore, Figures 2B.1 through 2B.6 show the comparison between characteristics of Wolfcamp and those of major shale plays. Figure 2B.1 shows that Wolfcamp's thickness is not close to that of any of the major shale plays. However, the TOC of Wolfcamp is very similar to four of the major shale plays, including Horn River, Ohio, Haynesville, and Marcellus.

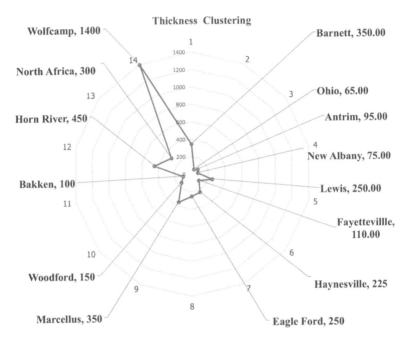

FIGURE 2B.1
Checking thickness clustering.

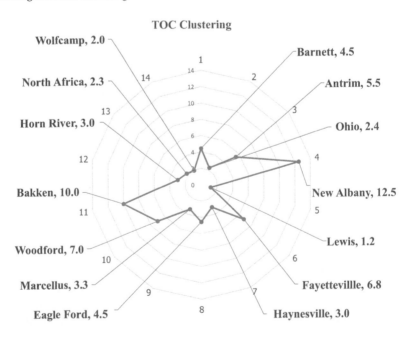

FIGURE 2B.2
Checking TOC clustering.

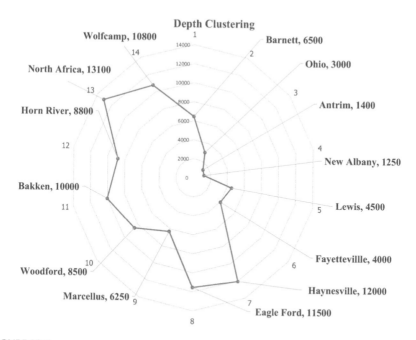

FIGURE 2B.3
Checking depth clustering.

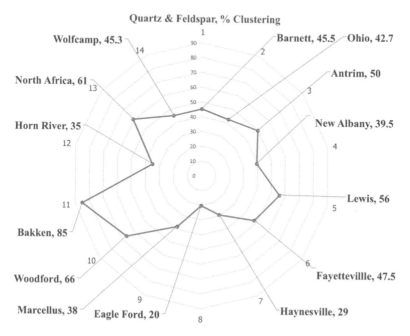

FIGURE 2B.4
Checking quartz wt. % clustering.

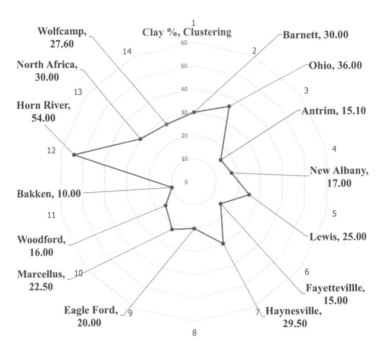

FIGURE 2B.5
Checking clay wt. % clustering.

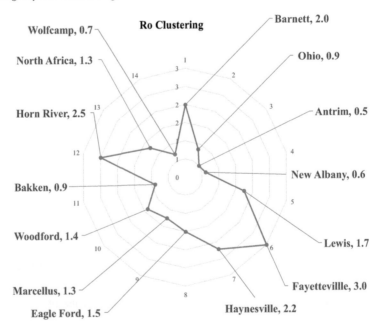

FIGURE 2B.6
Checking maturity clustering.

Appendix C: Completion Strategies of Shale Plays

TABLE 2C.1

Completion Strategy of Shale Plays

No.	Shale Play	Configuration of Horizontal Wells	Primary Completion Style	Fracture Design
1	Antrim	N/A	N/A	N/A
2	Bakken	Single lateral, multilateral	Barefoot open hole	Slick water/gel
			Nonisolated uncemented preperforated liner	PnP
			Fracture ports/ ball-activated sleeves	Fracture ports/ball-activated sleeves
			PnP[a]	100 mesh, 40/70, 30/50, 20/40, 16/20, and 12/18
3	Barnett	N/A	Cased and cemented, PnP completion	10%–12 fracturing fluid as a pad, 75%–85% as a sand-laden slurry mixture
4	Eagle Ford	N/A	Cased and cemented, PnP completion	Hybrid/water fractures
5	Fayetteville	N/A	Cased and cemented, PnP completion	Water fractures
6	Haynesville	N/A	Cased and cemented, PnP completion	Hybrid/water fractures
7	Horn River	N/A	PnP	Slickwater, 15 stages, 200 tons/stage, 17.6 Mbbl/stage
8	Lewis	N/A	N/A	N/A
9	Marcellus	N/A	Cased and cemented, PnP completion	Water fractures
10	New Albany	N/A	N/A	N/A
11	Ohio	N/A	N/A	N/A
12	Woodford	N/A	Cased and cemented, PnP completion	Hybrid/water fractures

[a] PnP = Plug and Perf.

TABLE 2C.2

Detailed Completion Treatment for Each Shale Play

	Major Play	Shale Play	Average Rate	Primary Fracture Type	Average No. of Stages	Completion	Primary Proppant Mesh	Hydrocarbon Type
1		Antrim	20	Xlink	12	PnP	N/A	Gas
2	Yes	Avalon/Bone Springs	70	WF/hybrid	18	PnP	30/50, 20/40	Oil/gas
3	Yes	Bakken	30	Xlink/hybrid	30	Fracture valves/PnP	30/50, 40/70, 20/40	Oil
4	Yes	Barnett	65	WF	9	PnP	100 mesh, 40/70	Gas
5		Cline	80	WF	25	PnP	100 mesh, 40/70, 30/50	N/A
6	Yes	Eagleford	50	Hybrid	16	PnP	30/50, 40/70, 20/40	Oil/gas
7	Yes	Fayetteville	75	WF	35	PnP	30/50, 40/70	Gas
8	Yes	Haynesville	75	Hybrid	15	PnP	30/50, 40/70	Gas
9	Yes	Horn River	70	WF/hybrid	18	N/A	30/50, 20/40	N/A
10	–	Lewis	–	WF	N/A	N/A	N/A	N/A
11	–	Mancos	50	WF/hybrid	17	PnP	20/40, 30/50, 40/70	N/A
12	Yes	Marcellus	85	WF	30	PnP	30/50, 40/70	Gas
13	Yes	Montney	35	WF/hybrid	25	PnP	30/50, 40/70, 20/40	Oil/gas
14	–	Mowry	40	WF/hybrid	14	PnP	30/50, 40/70, 20/40	N/A
15	–	New Albany	40	Hybrid	12	PnP	30/50, 40/70, 20/40	N/A
	–	New Albany	35	WF	N/A	Fracture valves/PnP	30/50, 20/40	N/A
16	Yes	Niobrara	45	Hybrid	30	Fracture valves/PnP	30/50, 40/70, 20/40	Oil/gas
17		Ohio	30	WF/hybrid	15	PnP	30/50	N/A
18		Utica/PP	80	WF/hybrid	35	PnP	30/50, 40/70, 20/40	Oil/Gas
19	Yes	Wolfcamp	75	WF	25	PnP	100 mesh, 40/70, 30/50	Oil/Gas
20	Yes	Woodford	85	WF	10	PnP	30/50, 40/70	Gas

TABLE 2C.3

Completion Treatment for Each Shale Play

Basin	Anadarko	Bakken	Barnett	DJ Basin	Eagle Ford	Haynesville	Marcellus
% Horizontal wells	80	94	89	77	96	72	83
HHP per stage	40k	30k	35k	30k	35k	60k	40k
Stages per well	20	30	12	20	25	20	20
Stages per day	8	10	3	9	6	4	6
Frac type	SW	XL	SW	SW/XL	SW/XL	SW/XL	SW
Avg. water requirement	2752 k	2245 k	3649 k	2876 k	4489 k	5644 k	4218 k
Depth	8428	10,270	7049	6909	9846	11,519	6993
Proppant type	20/40	20/40	40/70	100	30/50 and 20/40	40/70	40/70

TABLE 2C.4

Example of Stimulation and Completion Practices

	Woodford	Barnett	Haynesville	Marcellus	Eagleford	Bakken
Stimulation	Water frac/linear gel fluid system Rate to 60–120 BPM Stage size = 15,500 BBL/stage Prop volume = 4300 sks/stage Prop type: 10% 100 mesh, 90% 40/70 End concentration = 2 ppg Treating pressure = 6000–12,000 psi 6–12 stages	Water frac fluid system Limit rate to 30–100 BPM Stage size = 15,500 BBL/stage Prop volume = 4300 sks/stage Prop type: 60% 100 mesh, 40% 40/70 End concentration = 2 ppg Treating pressure = 4000–8000 psi 4–8 stages	Hybrid or WF fluid system 10–25 BPM per cluster (max rate = 80 BPM) 15 k lbs 100 mesh/sinterblast 85 k lbs 40/70 RCP 30/60 ISP 100 k gals/cluster XL fluid preferred 3–4 ppg/WF 2–3 ppg Treating pressure <12,000 ps	WF fluid system Rate = 60–100 bpm Acid = 2–5 kgal/stage Water = 300–600 kgal/stage Prop = 300–600 klb/stage Prop = 100 mesh, 40/70, 30/50, 20/40 3–4 ppg max PAD = 5%–15% 5–30+ stages, avg. 12–14	Hybrid/conv fluid system Rate = 40–60 BPM 12–15 stages 250–300 ft/stage 3–4 MM lbs prop (total) 30/60, 20/40 3–4 ppg 4 MM Gal 25,000 HHP	Hybrid/conv system Silverstim or hybor Gel frac fluid sys H Breaker-ViCon HT Rates = 20–40 BPM PDL 20/40 sand 20/40 ceramic 1–8 ppg for Xlink
Completion—Horizontal (best practices)	Cement casing More stages & shorter spans 6–12 stages/400-ft span 2–4-ft clusters, 4 per stage 6 jspf, rate ~+1.0 bpm/perf No bottom frac barrier	Cement casing More stages & shorter spans 4–8 stages/400-ft span 2–4-ft clusters, 4 per stage 6 jspf, rate ~+1.0 bpm/perf No bottom frac barrier	Cement casing 4500 ft lat/300-ft span TCP 1st stage Pump-down plug & gun perf 12–18 stages 24 hr operation 4–6 clusters of 1–2 ft All phasing & SPF	Cement casing Stage length = 150–450 ft Plug & perf method 4–6 JSPF EHD = 0.39″–0.47″ 1–5 clusters 2–4 ft each 30–120 ft apart	Cement casing 200–300 ft comp inter/stage 4–8 clusters 30–50 feet apart 1–2-ft length, 6 JSPF, 60° >2 BPM per perf (60 bpm thru 25 holes) 0.34″–0.38″ perf diameter DP & BIG hole charges	Open hole uncemented casing DSS/SP/VFLH Laterals drilled in 640 acres or 1280 acres 20–40 compartments Compartments = 200–400 ft

TABLE 2C.5

Well Production from the Same Area and Same Operator Using Two Different Completion Strategies

	Using the Recommended Completion Strategy	Using Other Completion Strategy
Initial oil rate, BOPD	598	405
No. of stages	15	32
Fluid type	Water frack and/or hybrid	Slickwater
Completion type	Perf and plug	Perf and plug
Primary mesh	40/70	40/70 Ottawa sand, 20/40 Ottawa sand
Average fresh water, gallons	2–4 million	2.3–4.2 million

Multiple Interval Fracturing

Colied Tubing	**Sliding Sleeves**	**Wireline**
Straddle Packer	Mechanical Shift	Limited Entry Stage Frac
Annualar Path with Packer	Ball Activated	Perf &Plug
Annuar Path with Sand Plugs	Multiple Ball Activated	Just-In-TIme Perforating
Annular Path with Anchor	Pressure Balanced with Packer	Ball Sealer or Particulate Diversion

FIGURE 2C.1
Cased hole completion categories and fracturing methods. (After Soliman, M.Y., Dusterhoft, R. 2016. *Fracturing Horizontal Wells*. First Edition, McGraw-Hill Education, N.p. Web.)

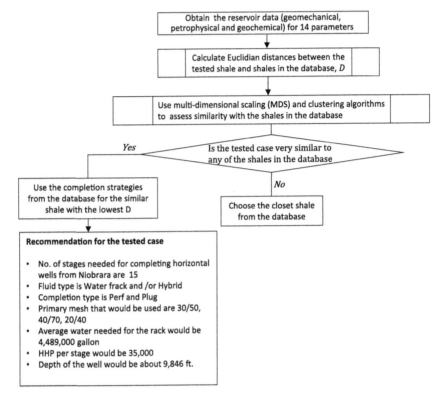

FIGURE 2C.2
Flowchart of implementing the introduced concept on the Niobrara shale horizontal well.

Appendix D: Details of the Computation of Euclidean Distances between Shale Plays

We illustrate this three-step procedure for the pair of shales Antrim & Bakken.

Step 1: Get means and standard deviations for each of the 14 parameters, obtained by averaging across the 12 shales. The data are in Table 2D.1.

Step 2: Compute standardized values of the 14 parameters (whose unstandardized values appear in Table 2.2).

Let (x_1, y_1) be a pair of (standardized) TOC values for (Antrim, Bakken), respectively:

$$x_1 = \frac{(5.50 - 5.2808)}{3.3309} = 0.0658$$

$$y_1 = \frac{(10.00 - 5.2808)}{3.3309} = 1.4168$$

TABLE 2D.1

Means and Standard Deviations for Each of the 14 Parameters

Parameter	Mean	Standard Deviations
TOC	5.2808	3.330868e + 00
Ro	1.5366	7.802369e − 01
Porosity	6.2458	2.695236e + 00
h	205.8333	1.272762e + 02
Depth	6475.0000	3.728484e + 03
Adsorbed gas	47.0277	1.654988e + 01
Gas content	142.1368	8.695543e + 01
Quartz	47.6256	1.075803e + 01
Clay	23.67565	1.225572e + 01
Pyrite	6.7563	9.410733e − 01
Calcite	19.9077	1.904221e + 01
Kerogen	7.76504	2.328932e + 00
E, million psi	3.4083	1.177827e + 00
v	0.2245	2.461996e − 02

and so on until we arrive at (x_{14}, y_{14}), the pair of standardized v values for (Antrim, Bakken):

$$x_{14} = \frac{(0.2066 - 0.2246)}{2.4620e - 02} = -0.7311$$

$$y_{14} = \frac{(0.2200 - 0.2246)}{2.4620e - 02} = -0.1868$$

Step 3: Finally, compute the Euclidean distance (d) between the two shale plays:

$$d = \sqrt{\begin{array}{l}(x_1 - y_1)^2 + (x_2 - y_2)^2 + (x_3 - y_3)^2 + (x_4 - y_4)^2 + (x_5 - y_5)^2 + (x_6 - y_6)^2 \\ + (x_7 - y_7)^2 + (x_8 - y_8)^2 + (x_9 - y_9)^2 + (x_{10} - y_{10})^2 + (x_{11} - y_{11})^2 \\ + (x_{12} - y_{12})^2 + (x_{13} - y_{13})^2 + (x_{14} - y_{14})^2\end{array}}$$

We obtain:

$$d = \sqrt{((0.0658 - 1.4168)^2 + \cdots + (-0.7311 + 0.1868)^2)} = 4.8307$$

3

Fracturability Index Maps for Fracture Placement in Shale Plays

3.1 Introduction

In hydraulic fracturing, the optimum spacing is a function of fluid flow and stress interference. The current trend in the industry is to place wells in a uniform manner and the fractures in an equally spaced distribution at the same time along the well path. Depending on the time span of fracturing operations, the net pressure created as a result of introducing a fracture will affect the initiation of subsequent fractures, which sometimes leads to reorientation of the created fracture and a nonplanar path fracture. Therefore, some fracture stages may not be as successful as planned.

The key to successful fracturing treatment in shale formations is the identification of "sweet spots." Productive shale consists of higher brittleness and segments that are easier to fracture. It may also include more quartz, feldspar, or carbonate and more organic matter than does less productive shale. Thus, mapping the best zones to fracture is usually a challenging process.

Multistage fracturing has become the preferred approach to the fracturing of shale reservoirs. However, it appears that in many cases, some stages or clusters do not contribute to production. Miller et al. (2011) analyzed more than 100 horizontal shale wells in multiple basins and found that two-thirds of total production comes from only one-third of the perforation clusters. They also reported that one-third of all perforation clusters were not contributing to production.

Petrophysical and geomechanical properties are the key to fracturing design. Miller et al. (2011) recommend use of reservoir and completion quality in designing stages and clusters along horizontal wells. Reservoir quality is defined by petrophysical properties of organic shale, which makes RQ a variable for development, such as maturity, porosity, and organic content. Completion quality is defined by the geomechanical parameters that are required to effectively stimulate the shale, such as stresses, mineralogy, and orientation of natural fractures. Currently, the shale brittleness indicator is used to identify a brittle and productive shale; see Jarvie et al. (2007) and Rickman et al. (2008).

Rickman et al. (2008) define brittleness as a function of Young's modulus and Poisson's ratio. They relate the success of any fracturing placement to the geochemical analysis that can be answered through petrophysical and lab measurements. Their work is more closely related to overall productive shale assessment, while our work focuses on productive segments of horizontal wells.

This chapter introduces new optimization criteria that would speed up the development process in unconventional reservoirs while reducing uncertainty and cost. Due to the expected severe heterogeneity of shale, methods such as the technique developed by Vasantharajan and Cullick (1997) would be inapplicable. Stegent et al. (2012) use a combined elevated factor vanadium as an indicator of total organic carbon and relative brittleness index for selecting intervals at which it would be easy to initiate fractures. They applied their technique for optimizing shale oil from the Eagle Ford formation.

3.2 Brittleness Index versus Mineralogical Index

Brittleness index is one of the main parameters for screening shale systems. It is a function of mineral composition and digenesis (Wang and Gale 2009). There is no general agreement regarding the definition of brittleness. Kahraman and Altindag (2004) and Altindag (2003) define brittleness based on tensile and compressive strength of the rock. Bowker (2007) suggests that the Barnett shale must contain less than 50% clay to be successfully fractured. Jin et al. (2014) redefine brittleness after Rickman et al. (2008). Kowalska et al. (2013) define brittleness as the sum of quartz and feldspar.

The most common brittleness indicators are given in the following equations:

$$E_n = \frac{E-1}{E_b-1} \times 100 \tag{3.1}$$

$$\nu_n = \frac{\nu_n-0.4}{\nu_b-0.4} \times 100 \tag{3.2}$$

$$BI = \frac{E_n+\nu_n}{2} \tag{3.3}$$

where:
  BI is the brittleness index based on mechanical properties.
  E is Young's modulus in millions psi.
  ν_n is Poisson's ratio.
  E_b and ν_b are correlation-based Young's modulus and Poisson's ratio of
    8 million-psi and 0.15, respectively.

BI based on mineralogy is written as given in Equation 3.4

$$BI = \frac{Q}{Q + C + CL} \tag{3.4}$$

where:

- Q = Quartz wt. %
- C = Carbonate wt. %
- CL = Clay wt. %

3.2.1 Isotropic versus Anisotropic Brittleness Index

Brittleness may vary horizontally and vertically. We believe that there is a 3D distribution of brittleness under certain applicable ranges. Based on an experiment by Britt and Schoeffler (2009), many prospective shale core samples with Young's moduli in excess of 3.5×10^6 psi exhibit some shear anisotropy in the core plug level. Similarly, Poisson's ratio may also have anisotropic property distribution.

3.2.2 Fracturability Index

Jin et al. (2014) introduced three fracturability index models. The first is the average product of normalized brittleness and strain energy release rate, the second is the average between normalized brittleness and normalized fracture toughness, and the last is the average of normalized brittleness and Young's modulus. Their range of fracturability indices varies between 0.4 and 0.84. Mullen and Enderlin (2012) introduced a complex fracturability index. The primary rock property input in their work is brittleness, which is directly correlated with Brinell hardness.

3.2.3 Objectives of This Work

- Develop a realistic fracturability index that may be used to design horizontal well paths.
- Design an optimum well spacing and fracturing scheme.
- Develop a fracture scheduling approach.

To achieve those objectives, the following limits have been set on the developed algorithm:

- It is assumed that the normal stress regime is the prevailing regime. Consequently, wells are drilled horizontally in the direction of minimum stress to create transverse fractures to maximize the drainage area created by the transverse fractures and the possibility of fracture tortuosity.

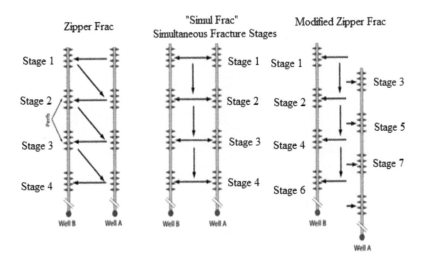

FIGURE 3.1
Different hydraulic fracture designs. (Modified after Rafiee, M., Soliman, M.Y., Pirayesh, E. 2012. Hydraulic Fracturing Design and Optimization: A Modification to Zipper Frac. *Presented at the Eastern Regional Meeting*, Kentucky.)

- The optimization algorithm allows wells to be placed within 30 degrees of the direction of minimum horizontal stress.
- Fractures created simultaneously, or nearly simultaneously, will be affected by stress shadowing.
- Recent approaches of hydraulic fracture design may be applied. Examples of these techniques are zipper, modified zipper, and simultaneous hydraulic fracturing, as shown in Figure 3.1.

3.3 New Fracturability Indices

3.3.1 Geomechanical Fracturability Index

To help guide well placement and fracture design optimization, two new correlations have been developed. This study finds that considering the gamma ray log improves the correlation coefficient of the selected model. The data used for building this model came from Permian Basin Wolfcamp shale. The first new correlation is given in the equation below, while the range of use is given in Table 3.1.

$$FI = 0.871 \times E'_n - (1/80)$$

(3.5)

TABLE 3.1

Data Range Used for Building FI Correlation

Parameters	Min. Value	Max. Value
E, psi	0.38 E6	9.75 E6
ν, ratio	0.02	0.38
Calcite, wt %	0.00	83.0
Quartz, wt %	6.00	75.0
Pyrite, wt %	0.00	8.00
Clay, wt %	3.00	49.0
Density, ρ, g/cc.	2.40	2.71

where:

$$E'_n = \frac{[(E/(1-\nu^2))-(E/(1-\nu^2))_{min}]}{[(E/(1-\nu^2))_{max}-(E/(1-\nu^2))_{min}]} \qquad (3.6)$$

where:

$(E/(1-\nu^2))_{min}$ = Minimum plane strain Young's modulus in the reservoir
$(E/(1-\nu^2))_{max}$ = Maximum plane strain Young's modulus in the reservoir

FI accounts for the effects of mineralogy and energy release for creating fractures.

3.3.2 Resistivity Fracturability Index

Formation resistivity is a sensitive parameter controlled by the insulating properties of solid matrix materials and the conductive paths available only where there are interconnected pores with an ionic conductance pathway. In conventional reservoirs, porous water-bearing intervals are conductive and nonporous intervals resistive. In unconventional reservoirs, high kerogen content organic shales are resistive and ductile high clay content shales are conductive. Formation resistivity is thus an indicator of frack barriers where resistivity is low and an indicator of harder, low-clay, high-TOC organic shales that are brittle and hence more easily fracked.

An interesting consequence of the Archie relationship is that the reciprocal square root of the formation resistivity is linear with the formation's bulk volume of water. In other words, the bulk volume of water, given by the product of porosity and water saturation (φ Sw), when squared is proportional to the rock conductivity. This concept is the basis for the well-known Hingle plot used by log analysts and, indirectly, is also used in the Passey overlay as a means to determine TOC from sonic and resistivity logs. The second proposed fracturability index is based on this sidebar to the Archie

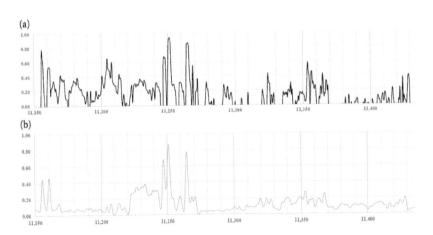

FIGURE 3.2
(a) Resistivity fracturability index derived exclusively from formation resistivity.
(b) Fracturability index based on geomechanical data.

relationship. If $R' = 1/\sqrt{R_t}$, then the resistivity fracturability index (RFI) is given by Equation 3.7:

$$RFI = aR' + b \tag{3.7}$$

where a and b are characteristic "tuning" parameters for local conditions. The RFI has the potential to be tailored not only to particular plays (Eagle Ford, Haynesville, etc.) but within any one play, the values of a and b can be fine-tuned to imitate any of a number of the other indicators that have been published in the literature, be they based on mineralogy or formation elastic properties. Thus, a family of resistivity-based RFIs could be published to fill in gaps in a field caused by a lack of a full logging suite. An experienced petrophysicist will be able to imitate both the mineralogical and geomechanical indices.

For demonstration purposes, Figure 3.2a and b shows how the resistivity index compares with a geomechanical fracturability index that requires log inputs of neutron, density, and array sonic measurements.

3.4 Optimization of Number of Wells and Fractures in a Reservoir

The final objective of this chapter is to present an approach to optimize the net present value of a fracturing project in a shale play. This approach requires the design of the number of wells, fractures, productivity, and so on. In the next section, the various parameters used in designing the optimization process are discussed. In the optimization work presented in this chapter, we assume that the wells are horizontal or very close to horizontal.

3.5 Formulating the Optimization Approach

In order to arrive at the proper formulation, the following parameters are determined by use of the optimization technique:

- Number of fractures
- Fracture dimensions
- Fracture locations
- Optimum scheduling

The computational approach to FI calculation is based on mathematical optimization using integer programming. AlQahtani et al. (2013) have recently proven integer programming's superior performance in vertical well placement relative to other optimization techniques. The objective function given in Equation 3.8 states that the goal of the developed scheme is to maximize the sum of the fracturability indices at the points where fractures propagate.

$$\text{Maximize} \sum_{(i,j)\in I\times J} FI_{(i,j)}x_{(i,j)} \tag{3.8}$$

The maximization of the function in Equation 3.8 is subject to constraints in the following equations:

$$x_{i_1,j_1} + x_{i_2,j_1} \leq 1, \quad \left| \frac{(i_1,j_1),(i_2,j_2)}{\sqrt{(i_1-i_2)^2+(j_1-j_2)^2}} < D, (i_1,j_1),(i_1,j_2) \in I\times J \right. \tag{3.9}$$

Equation 3.9 states that a minimum spacing between adjacent fractures is specified.

$$\sum_{(i,j)\in I\times J} x_{i,j} \leq N \tag{3.10}$$

Equation 3.10 shows constraints of the total number of fractures per given shale plane.

$$x_{(i,j)}\{0,1\} \tag{3.11}$$

Equation 3.11 states that the location is a variable $x(i, j)$ that represents the coordinates (i, j) in the shale formation plane and is equal to 0 if the location is not chosen, or 1 otherwise.

3.5.1 Additional Design Constraints

Net pressure < preset value based on the shale play and planned fracture design.

- Completion spacing constraints
- Angular deviation constraints
    - Well azimuth within 30 degrees of the direction of minimum stress
    - Well inclination <= 90; inclination is usually around 90
- Maximum length constraints
    - Fracture length, X_f < distance to drainage boundary X_e

Fracturability index is a continuous range: $[0.....1] = f(E, v)$.
The best well path achieves the condition in Equation 3.12. Equation 3.12 indicates that the well path will reach the maximum fracturability index. The new fracturability index is used as an input matrix for well and fracture placement.

$$\sum_{k=0}^{n} \binom{n}{k} \quad \text{max average } FI(X,Y,Z) \tag{3.12}$$

The flowchart in Figure 3.3 depicts the steps of a method of generating a fracturability index map that will be used in the optimization process. Once the process is initiated, then, as shown above, a step may be implemented in which the well path is divided into segments. Thereafter, the step may be implemented for identifying the order of fracture locations along the well path. Such ordering may include the best fracture from a production point of view, then a second fracture stage, and so on. Then, as depicted, an operation with the same approach for a second well is repeated, as well as for subsequent wells, until the simulation runs determine the prioritized segments.

In conventional fracturing design, implementation of a fracture may affect the initiation of the subsequent fractures, which may sometimes lead to

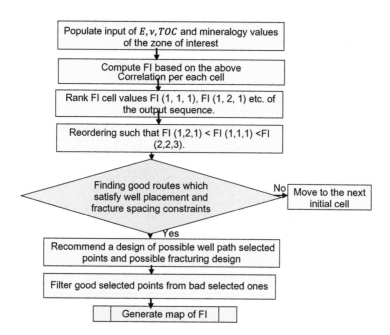

FIGURE 3.3
Flowchart used in optimization.

reorientation of the fracture and obtaining a nonplanar fracture path. The reorientation will affect the effective fracture width and production efficiency per stage. Introduction of a new a design based on the developed index may solve this problem.

3.6 Method of Solution

The optimization problem can be solved by the use of integer programming. A commercial solver (GUROBI) can help find the optimum solutions. To solve the examples given in this chapter with the IP approach, a solver based on branch and cut that can handle large-scale optimization problems was used. For details on branch and cut, refer to Nemhauser and Wolsey (1999).

The computations were carried out on a single high-speed machine. For tests using IP, the formulations generated were written in VBA and the optimization models were resolved using the GUROBI 5.6 solver (GUROBI is a state-of-the-art professional solver package) with default settings set to the branch-and-cut solution strategy and absolute gap tolerances set to zero. The other controls are the same as those described in AlQahtani et al. (2013).

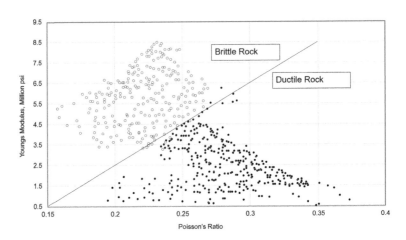

FIGURE 3.4
Brittleness index parameters after Rickman et al. (2008).

3.7 Case Study 1

Rickman et al. (2008) developed the first correlation used by the industry for differentiating between ductile and brittle behavior. Figure 3.4 shows a plot of the data used by Rickman et al., while their range of data is listed in Table 3.2. Analyzing Rickman et al.'s data points using the newly proposed technique, we obtained agreement between their brittleness index criteria and our newly developed FI correlation criteria. Table 3.3 shows this agreement. Using the model developed in this chapter, the data reported by Rickman et al. were replotted in Figure 3.5. The results show a well-defined trend and fairly straightforward definition of brittle and ductile behavior.

3.7.1 Summary of Correlations

Table 3.4 gives a summary of recommended fracturability indices and suggested classification. Table 3.4 shows different tested ranges of mineralogy indices versus their corresponding ranges of FI. Comparing the results from the new fracturability index versus the results reported by Rickman et al. (2008), we can conclude that the cutoff between ductile and brittle behavior is in the range of 0.33–0.65.

3.8 Case Study 2 (Well Placement Case Study)

The first case is a shale reservoir with populated Young's modulus and Poisson's ratio for each cell. We started the process of calculating FI as a

TABLE 3.2

Data Range Used for Developing Rickman et al.'s
Correlation

	ν	E, psi	$E/(1 - \nu^2)$, psi
Minimum value	0.16	0.5 E + 6	1.8 E + 6
Maximum value	0.37	8.5 E + 6	6.2 E + 6

TABLE 3.3

Testing This Chapter's FI against Selected Data of Rickman et al.'s
Correlation

ν	E	$E/(1 - \nu^2)$	Normalized $E/(1 - \nu^2)$	FI	
0.31	2.0 E + 6	2.2 E + 6	0.09	0.09	
0.34	1.6 E + 6	1.8 E + 6	0.00	0.00	
0.29	2.8 E + 6	3.0 E + 6	0.28	0.22	Ductile
0.28	3.0 E + 6	3.0 E + 6	0.33	0.29	
0.25	4.0 E + 6	4.0 E + 6	0.56	0.46	
0.19	6.0 E + 6	6.0 E + 6	1.00	0.90	
0.22	5.5 E + 6	5.7 E + 6	0.90	0.80	Brittle
0.23	5.3 E + 6	5.5 E + 6	0.85	0.75	

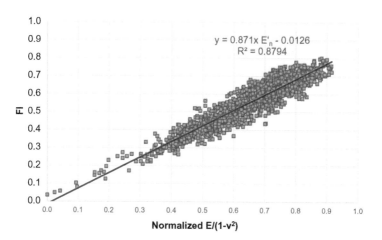

FIGURE 3.5
Combined ductile and brittle shale reservoir based on developed FI model.

property for each cell. The model consists of 370 × 22 cells with the following
characteristics (Table 3.5).

Figure 3.6 illustrates an example fracturability index map. A shale gas
model as recommended by the optimizer is shown in a 3D map, which can

TABLE 3.4

Recommended FI Ranges

Mineralogical Index (%)		Classification	FI—This Chapter	
70	80	Highly brittle	0.88	1.00
60	70	Very brittle	0.77	0.88
50	60	Brittle	0.66	0.77
40	50	Low brittleness	0.55	0.66
30	40	Transition	0.44	0.55
20	30	Ductile	0.33	0.44
10	20	Very ductile	0.22	0.33

be generated in accordance with the methodology described earlier in the chapter. Figure 3.6 illustrates the tested model of FI populations. The blue cells are suggested by the solver. These points are used as the initial points for wells and fracturing positioning. Based on these points and by use of the solver, the best 120 fracture locations were selected from 8140 possible fracture locations in the shale model.

3.8.1 Remarks on Case Study 2

- The optimizer selected 120 possible points for possible fracture locations distributed for three horizontal wells.
- Refracturing of this reservoir is recommended because the selected locations have fairly high fracturability indices of larger than 0.8.

TABLE 3.5

Reservoir Input Data Used in Well Placement Study

ΔX, ΔY, ft.	25,200
Lateral length, ft.	9600
Initial reservoir pressure, psi	5400
Pay zone, ft.	827
L_f, ft.	390
Fracture width (grid dimensions 3 × 3), in.	0.12
Depth to the top of the formation	6105
c_f, psi^{-1}	3E-06
Porosity, %	10
No. of fractures	120
Fracture spacing	80 ft. (min)
Reservoir lifetime = (7305 days) based on (BHP = 1500 psia and Qg = 30 Mscf/d)	20 years

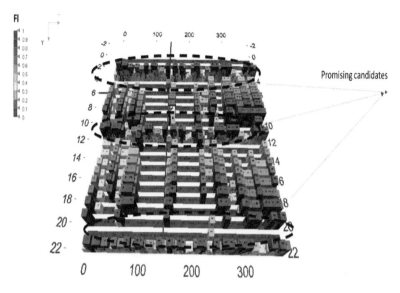

FIGURE 3.6
Plot of FI distribution for shale layer.

3.9 Case Study 3 (Fracture Placement Case Study)

This case study was originally presented by Jin et al. (2014). According to Jin et al.'s approach, the light red shaded sections in Figure 3.7 are candidates for hydraulic fracturing. This well is located in Marble falls, Upper Barnett,

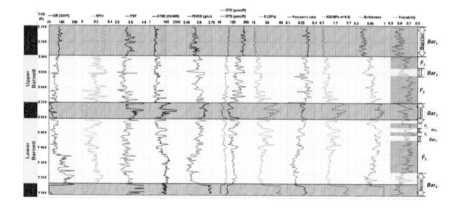

FIGURE 3.7
Screening hydraulic fracturing candidates with their fracturability index. (From Jin, X. et al. 2014. Fracability Evaluation in Shale Reservoirs—An Integrated Petrophysics and Geomechanics Approach. *Presented at SPE Hydraulic Fracturing Technology Conference* The Woodlands, TX, February 4–6.)

TABLE 3.6

Mechanical Properties of Data in Figure 3.7

	E, psi	ν	$E/(1-\nu^2)$
Maximum value	15,154,379	0.38	16,429,417
Minimum value	2,615,066	0.16	2,866,150

TABLE 3.7

Calculations of FI Using Approach Presented in This Chapter

Depth, ft.	Normalized $E/(1-\nu^2)$	FI
203	1.00	0.86
309	0.98	0.84
402	0.97	0.83
504	0.98	0.84
604	0.96	0.83
705	0.54	0.46
809	0.86	0.74
900	0.75	0.64
1004	0.84	0.72
1099	0.79	0.68
1199	0.84	0.72
1300	0.00	0.00

Forestburg Limestone, Lower Barnett, and Viola Limestone. Table 3.6 gives input ranges of data used in their work. Table 3.7 shows the calculated fracturability indices for the cross-section of data in Figure 3.7.

There is a fairly good match between what Jin et al. have suggested and the technique developed in this chapter. Both techniques suggest fracturing brittle rocks and eliminating thin sections of high FI.

3.9.1 Conclusions

1. New fracturability indices based on mechanical formation properties and resistivity measurements have been developed. These new fracturability indices identify the fracturing candidates more clearly than the existing indices.
2. The chapter presents a new optimization technique for placing wells and fractures.
3. The validity of the developed techniques has been illustrated through the use of field examples.

Appendices

Appendix A: Abbreviations

E'_n	Normalized plane strain modulus, million psi
p_{net}	Net pressure, psi
D_{min}	Minimum well spacing, ft.
L_f	Fracture half-length, ft.
X_e, Y_e	Rectangular reservoir shape dimensions in x and y directions, ft.
$x_{(i,j)}$	X_Y location of a fracture in the reservoir represents the location (i, j) in the shale formation grid
$\Delta X, \Delta Y$	Model grid dimensions in x and y directions, ft.
BHP	Bottomhole pressure, psi
E	Young's modulus, psi
E'	Plane strain modulus, million psi
F	Fracture stage
FI	Fracturability index
L	Well lateral length, ft.
Qg	Gas flow rate, mscf/d
W	Horizontal well
X	Coordinate axis along well path, ft.
Y	Coordinate axis along fracture path, ft.
ν	Poisson's ratio
ρ	Density, lb./ft³

Appendix B: Summary of Fracturability Indices

Brittleness Index Based on Mechanical Properties

Rickman et al. (2008)

$$E_{n\,britt} = \frac{E-1}{8-1}*100$$

$$\nu_{n\,britt} = \frac{\nu_n - 0.4}{0.15 - 0.4}*100$$

$$BI = \frac{E_{n\,britt} + \nu_{n\,britt}}{2}$$

Gray et al. (2012)

$$BI = 50\% * \left(\frac{E - E_{min}}{E_{max} - E_{min}} + \frac{\nu - \nu_{max}}{\nu_{min} - \nu_{max}} \right)$$

Jin et al. (2014)

$$BI = \frac{E_n + \nu_n}{2} \quad E_n = \frac{E - E_{min}}{E_{max} - E_{min}} \quad \nu_n = \frac{\nu_{max} - \nu}{\nu_{max} - \nu_{min}}$$

Brittleness Index Based on Mineralogy

Brittleness index, Jarvie et al. (2007)

$$BI = \frac{Q}{Q+C+CL}$$

- Q: Quartz
- CL: Clay
- C: Carbonate

Brittleness index, Wang and Gale (2009)

$$BI = \frac{Q+\text{Dol}}{Q+\text{Dol}+\text{Lm}+CL+TOC}$$

- Q: Quartz
- CL: Clay
- C: Carbonate
- Dol: Dolomite
- TOC: Total organic carbon

Bowker (2007)

Mineralogical index $= \Sigma$Quartz

Kowalska et al. (2013)

$BI = \Sigma$Quartz + feldspar

Brittleness Index Based on Failure Criteria

Altindag (2003)

$$BI = \frac{\sigma_c * \sigma_t}{2}$$

where σ_c is the compressive strength and σ_t is tensile strength.

Kahraman and Altindag (2004)

$$BI = \frac{\sigma_c - \sigma_t}{\sigma_c + \sigma_t}$$

Fracturability Index

Jin et al. (2014)

$$FI = \frac{B_n + G_{C\_n}}{2}$$

$$FI = \frac{B_n + K_{IC\_n}}{2}$$

$$FI = \frac{B_n + E_n}{2}$$

where B_n, G_{C_n}, K_{IC_n}, and E_n are normalized values of brittleness, strain energy release rate, fracture toughness, and Young's modulus.

4

Is Fracturability Index a Mineralogical Index? A New Approach for Fracturing Decisions

4.1 Introduction

With the increased demand on oil and gas resources from shale plays, it has become profitable to develop these resources. Many successful shale plays, such as Bakken, Eagle Ford, and Barnett, have been developed through horizontal wells and multistage fracturing. The recent big shale resources added to these common successful plays are Wolfcamp across Permian Basin in Midland (West Texas). Wolfcamp formation lies beneath Spraberry and Dean formation. Over the last 20 years, companies have been developing more than 10 formations using vertical wells. The Wolfcamp is over 1000 feet, subdivided into A, B, C, and D. These divisions are principally shale rich, with the upper part more carbonate rich. The key is placing wells in sand-rich intervals with enough kerogen.

In the case of an unconventional gas and oil reservoir for which gas and oil production totals over the entire reservoir area are given (or estimated), we are interested in the well placement problem, that is, maximizing the total gas and oil production of the reservoir while minimizing the cost and considering certain constraints such as number of nonconventional wells and distances between wells. Spacing between wells and fractures is also constrained by stress regime and a new MI. The new MI serves as a quality map to guide the selection of nonconventional wells and hydraulic fracture locations.

There is no doubt that horizontal drilling in the direction of the minimum stress regime and hydraulic fracturing of those wells are key factors in the revolution of shale gas and oil in the United States. The variation in shale organic properties, rock mechanical properties, and the nature of shale mineralogy can contribute to a new sweet-spot identification index that can be linked to a mixed integer programming–developed optimization algorithm.

Mathematical optimization using integer programming proved its superior performance in vertical well placement (for details on its performance, see AlQahtani et al. [2013]). IP guarantees optimality in any proven solution, unlike evolutionary metaheuristics (genetic algorithms [GAs]), which can give optimal solutions but is sometimes trapped in local optima. An integrated approach includes geomechanics, geochemistry, and petrophysics, considering that rock and fluid properties can be a superior tool in identifying shale gas reservoirs. The mathematical optimization in a form of IP is used to minimize the number of wells that maximize the quality points of MI. Optimizing the number of wells and the location of each deviated well in a Y-Z plane is an objective.

Representative maps, including and not limited to geochemical data (quartz and clays [illite, smectite, kaolinite, chlorite, kerogen, pyrite k-feldspar]), populated in a 3D reservoir model can give a quick idea about the most mineralogically brittle places in the 3D model. The second map could be a geomechanical data map based on geomechanics, while the third map is total organic carbon. Populating TOC can be accomplished by use of a commercial reservoir engineering tool. Other maps such as permeability, porosity, and maturation (Ro) can be used as a definition for sweet spots in shale gas reservoirs.

Coring, cuttings (mud logging), and seismic under a certain resolution can be used for building a reliable 3D dynamic model for a geochemical index. Microseismic can possibly be used to map out reactivated natural fractures near the well bore, and it can also be correlated and give an idea about the distribution of natural fractures within the reservoir, which can help in getting accurate geomechanical indices.

4.2 Background

The following correlations have been used to locate wells and fractures, and industry has used them as a proxy for mineralogical brittleness and fracturability. See Table 4.1.

A timeline of brittleness indicators used for locating wells and fracks in shale resources is shown in Figure 4.1.

4.3 Well Placement in Conventional Reservoirs

To find optimum well locations, algorithms were used to efficiently accelerate the process and automate the placement of wells. Many authors approached

TABLE 4.1

Common Mineralogical Brittleness Index Based on Mineralogy

Kowalska et al. (2013)	$BI = \sum Quartz + Feldspar$
Buller et al. (2010)	$RBI = \dfrac{Brittle\ Mineral\ Proxies}{Brittle + Ductile\ Mineral\ Proxies}$
	• RBI = Relative brittleness index
Wang and Gale (2009)	$BI = \dfrac{Q + Dol}{Q + Dol + Lim + CL + TOC}$
	• Q = Quartz
	• CL = Clay
	• C = Carbonate
	• Lim = Limestone
	• TOC = Total organic content
Bowker (2007)	Mineralogical Index = $\sum Quartz$
Jarvie et al. (2007)	$BI = \dfrac{Q}{Q + C + CL}$
	• Q = Quartz
	• CL = Clay
	• C = Carbonate

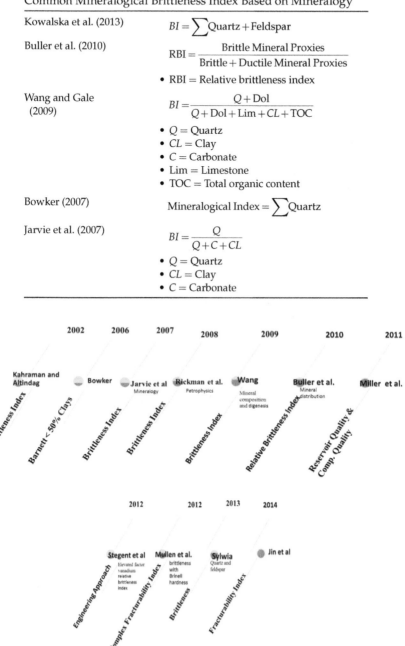

FIGURE 4.1

Common indicators used to differentiate brittle versus ductile rocks through literature.

the problem of well placement in conventional reservoirs by simulating vertical well placement and considering different spacing and certain geometrical constraints.

Vasantharajan and Cullick (1997) developed an approach to automate this selection by using mathematical optimization (integer programming) formulation and integer programming algorithms to suggest the best possible well locations in the quality maps for each grid, such as net pay. Such a quality map can be used as an input to well site selection. Da Cruz et al. (1999) introduced the concept of a quality map applied to well location selection and scheduling. It works for every grid by running the flow simulator with a single well and varying the location of the well in each run to have good convergence of the entire horizontal grid. A quality map uses static reservoir properties with each of the grid cell quality values; it is a 2D property representation of the reservoir in x and y. Nakajima and Schiozer (2003) also used quality maps for placing horizontal wells to optimize reservoir performance with horizontal wells. They developed a methodology by checking the performance of one well or group of wells and field. The aim was to determine parameters that affect horizontal well productivity, but their methodology does not give the optimum solution. Maschio et al. (2008) use a genetic algorithm and quality map for production strategy optimization, in other words, minimizing the number of wells. Their approach suggested a number of production/injection wells and production and injection flow rates.

4.4 Well Placement in Unconventional Reservoirs

Understanding shale reservoir quality helps in optimum placement of vertical or horizontal wells in shale. The performance of horizontal wells depends on placing these wells in the preferred target zone due to vertical heterogeneity, which tends to affect drilling and completion performance (Azike 2011). Research is done to place the laterals in the most productive shale (having more quartz content and less clay), which can guarantee the created fracture is optimum and more complex. Shebl et al. (2012) introduced a cross-plot approach of Young's modulus and Poisson's ratio to help place wells in brittle shale gas and oil, which is easier to fracture than ductile shale. That case study recommends placing wells in brittle shale and thin carbonate.

Wilson et al. (2012) presented a general framework for applying optimization to the development of shale gas reservoirs. They used a direct search optimization approach to determine the optimal locations, lengths, and number of fracture stages for a set of horizontal wells. Cheng et al. (2009) presented a methodology of optimization of infill wells in tight gas reservoirs using a sequential inversion algorithm for rapid history matching and successive selection strategy for infill candidate well locations.

Our procedure is as follows. Using well data analysis to build a correlation of the mineral composition of a shale reservoir, the built model is defined by grid blocks of dimensions $125 \times 125 \times 5$ ft. Values of porosity and permeability were obtained from the core data report, then assigned to grids. The average reservoir permeability is 164 nanodarcies (nd).

The current work consists of five stages to complete a shale play field development, including well placement in the most mineralogical brittle rock: stage 1, extensive data analysis; stage 2, the building of a representative mineralogical scale based on mineralogy; stage 3, development of a 3D mineralogy model; stage 4, problem formulation; and stage 5, presentation of an optimization approach.

4.4.1 Stage 1: Data Analysis

Multiple wells from Permian Basin are analyzed to build a representative geochemistry map and mineralogical index. Petrophysical log data from two wells in Wolfcamp were used with geostatistical techniques to construct the detailed geological model used here for testing. Many relationships of parameters (such as porosity, permeability, quartz, clay content, E [Young's modulus]) were tested in this work to understand Wolfcamp shale. See Table 4.2.

4.4.2 Stage 2: Building Mineralogical Index

Mineralogy data points from Permian Shale Wolfcamp are used extensively to build the best representative correlation. The index may be used as a characterization of shale proxy or as a quality point in terms of organic

TABLE 4.2

Data Used for Building the Correlation

1806 Samples		
	Minimum Value	**Maximum Value**
Parameters		
Quartz, wt. %	6.00	75.0
Calcite, wt. %	0.00	84.0
Clay, wt. %	3.00	49.0
Pyrite, wt. %	0.00	8.00
Photoelectric index (P_e), barns/electron	2.61	5.71
Density, ρ_z, g/cc	2.41	2.71
h, ft.	900	
E, psi	0.38 E6	9.75 E6
υ, ratio	0.02	0.38
Calculated Parameters		
New mineralogical index	0	0.95

content or higher content of silica. The developed correlation is used to build a 3D shale lithofacies-based MI for Wolfcamp across Permian Basin.

After an extensive search for the best combination of representative parameters, the following combination represents the mineralogical index.

$$MI = 1.09 \times \frac{Quartz + Feldspar + Pyrite}{Quartz + Feldspar + Calcite + Clay + Pyrite} + \frac{1}{8.8} \quad (4.1)$$

The variable

$$\frac{Quartz + Feldspar + Pyrite}{Quartz + Feldspar + Calcite + Clay + Pyrite}$$

is the mineralogical brittleness index, and the input parameters in Equation 4.1 are in weight %.

MI is crucial because it works as a proxy for placing wells in the most productive shale portions. Organic siliceous shale has the highest quartz and kerogen content.

4.4.3 Stage 3: The Three-Dimensional Mineralogical Model

The main purpose of building a representative 3D model of shale mineralogy was to test the developed approach through a broad variety of shale architecture. Shale plays are not identical in terms of composition.

The deep understanding of the model works as a guide for future development of this resource. Future planning includes placing wells in the shale model and fractures along the selected well paths. The mineralogical index map is a valuable result of mineralogy analysis. Such a map can be a valuable way of presenting information that clearly indicates good sweet spots versus bad regions within the model. An optimization algorithm works for suggesting well placement based on detailed features of the model. Spacing and cost of placing deviated wells will work as a guide in suggesting the minimum number of required wells.

4.5 Unconventional Well Placement Problem Formulation

In order to get the proper formulation, certain questions must be addressed: How many deviated wells are needed for draining a certain reservoir? What is the optimum combination of deviated wells to drain a certain reservoir? What is the optimum well scheduling?

The use of integer programming to find the hot points that minimize risk factors and maximize reservoir quality points is presented in the following section.

We have two quality points:

1. Mineralogical index
2. Kerogen content (TOC distribution)

For the purpose of this chapter, we focus on geochemical analysis and kerogen content as a proxy for guiding the optimization algorithm.

4.5.1 Stage 4: Problem Formulation

For the problem definition, a detailed explanation of the formulation is listed in Appendix B.

In the optimization work presented in this chapter, we assume that the wells are horizontal or very close to horizontal.

Objective function

The objective function given in Equation 4.1 states that the goal of the developed scheme is to maximize the sum of the mineralogical indices for the well with the maximum profit.

$$\max \sum_k \sum_{i \in X^k} MI_i x_i^k \tag{4.2}$$

Assuming fractures are perpendicular to the wellbore:

Input parameters (Table 4.3)

Maximum length constraints

- Well length $L \leq$ drainage boundary Y_e
- Fracture length $X_f <$ distance to drainage boundary X_e

TABLE 4.3

Main Input Parameters to Optimization Model

Length of horizontal well, ft.	5000–15,000
Spacing between wells, ft.	300–700
Spacing between fractures, ft.	80–500
Spacing between fractures from neighboring wells, ft.	50–100

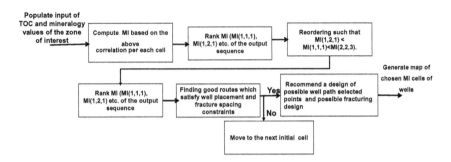

FIGURE 4.2
Well placement algorithm.

To maximize Equation 4.2 using IP methodology, we use a commercial solver based on branch-and-cut optimization. We have used the GUROBI software package as a solver to solve practical instances of the problem, and we have used VBA in generating instances of the problem.

4.5.2 Stage 5: Optimization Approach of Well Placement

We are dealing with the shale oil model, a single model consisting of many planes (Y-Z), each plane of which could be made a quality map by use of static reservoir properties with each grid cell's quality values. Our approach starts with a nontraditional method of placing wells within shale, a method that considers the static model from the Y-Z plane (one model may consist of many planes [Y-Z], depending on the number of grids in the plane, as well as the budget available for placing the wells). Respecting the fracture half-length, an initial number of wells is suggested with a proposed spacing. It is a 3D model whereby multiple planes form a 2D representation of the reservoir in the y-z plane. Then, there is a second plane of cells, until reaching the optimum path that connects all dots for different planes to give the optimum trajectory ending with an optimized well. The minimum stress direction is East-West.

Figure 4.2 shows the flowchart that summarizes the process of generating a mineralogical index map to be used in the optimization process.

4.6 Well and Fracture Placement Case Study Using Mathematical Optimization

A subsurface reservoir model was built using integrated data from cores and sonic data in order to build a representative dynamic model, including accurate facies distribution, mechanical properties, and petrophysical properties.

TABLE 4.4

Reservoir Input Data Used in Well Placement Study

ΔX, ΔY, ft.	125,125
Lateral length, ft.	10,000
Initial reservoir pressure, psi	3900
Well spacing, ft.	800
Fracture width (grid dimensions 3×3), in.	0.12
c_f, psi^{-1}	3.00E−06
Minimum fracture spacing	80 ft.
Pay zone, ft.	900
X_f, ft.	125–1250
Depth to the top of the formation, ft.	7900
Porosity, %	10
TOC, wt. %	2.3
RO, wt. %	1.3
Average permeability, nd	160

The data in Table 4.4 are a shale reservoir model with populated Young's modulus and Poisson's ratio per each cell. MI was first calculated as a property of each cell.

The simulation model consists of an 80×164 shale 2D quality map (see model parameters in Tables 4.2 and 4.4). In this characterization of wells and fracture types over shale derived from the built correlation of FI and plane strain Young's modulus, the red regions suggest where a horizontal well will form, the blue where the rock is more ductile and less likely to support fracturing, and the yellow where aligned and suboptimal fractures will form. The chosen layer is located at 8400 feet in a zone of the lowest minimum horizontal stress, higher organic matter content, and higher mineralogical index. Figure 4.3 shows the input matrix as a 2D map.

The model consists of $80 \times 80 \times 164$ cells with the following characteristics.

Figure 4.3 demonstrates a 2D map from the first slice of the well's entry to the reservoir. As shown in the figure, clay-rich layers are considered a hazard zone for placing horizontal wells. Figure 4.4 demonstrates the x-z plane with initial well placement suggested to pass through sweet spot locations.

Figure 4.5 shows a heat map of TOC population for a 2D shale map (Y-Z) plane; red cells show the sweet spot of the reservoir, which contains organic matter for placing wells.

Figure 4.6 shows a heat map of the MI population for a 2D shale map (Y-Z) plane; red cells show the sweet spot facies for placing wells. See Figure 4.7.

Figures 4.8 through 4.11 show a quartz and mineralogy mapping of the shale oil model over the entire depth.

Figure 4.8 shows, from left to right, Poisson's ratio, Young's modulus, TOC, mineralogy, FI (Alzahabi et al. 2015b), FI (Alzahabi et al. 2015b; normalized per zone), BI (Rickman et al. 2008), and MI (this chapter).

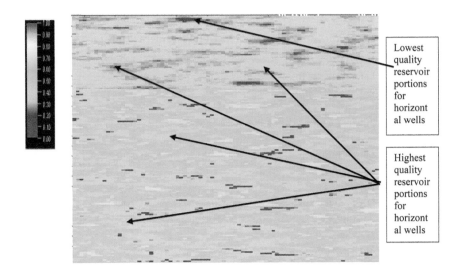

FIGURE 4.3
The first plane (slice) of MI distribution showing higher reservoir quality portions for initiating horizontal wells.

Figure 4.9 shows an example of locating wells and fractures for case study 1. For maximum fracture half-length of 17 cells, minimum fracture spacing of 5 cells, and minimum well spacing of 5 cells, the optimum solution is shown in Figure 4.9. For two wells to drain the reservoir, 32 fracture stages in a zipper-staggered form are recommended.

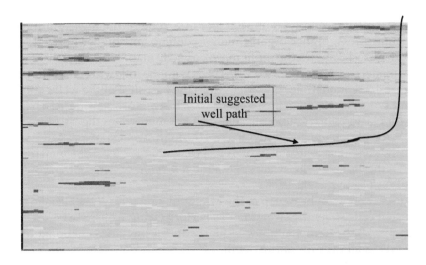

FIGURE 4.4
Typical optimized well path, middle plane (slice) of MI distribution showing higher reservoir quality portions for initiating horizontal wells.

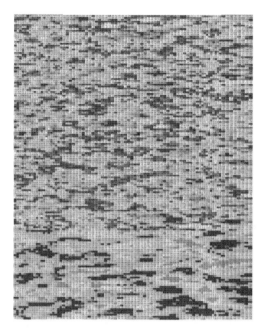

FIGURE 4.5
TOC map of the first entry plane to the shale model with TOC cells colored red to show the sweet spots.

FIGURE 4.6
MI map of the first entry plane of the model with cells colored red to show the sweet spots.

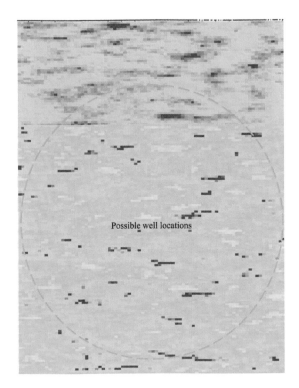

FIGURE 4.7
MI distribution for first slice east-west direction, minimum horizontal stress direction.

The placement of two wells, along with the recommendation for the number of fractures controlled by MI distribution in a shale oil model as recommended by the optimizer, is shown in a 2D map in Figure 4.9.

The sweet-spot allocation method is based on facies identification and TOC content for the guiding of well and fracturing placement. See Table 4.5.

The table confirms that Young's moduli and Poisson's ratios vary based on mineralogy composition within the rock. The mineralogical index has many forms. The MI introduced in this chapter depends on mineralogy variations within the reservoir.

The horizontal well placement entry locations yielded by the optimization approach for the shale model match the TOC with MI. There is a match between the two proxies used for placing the wells for the five cases shown.

4.7 Conclusions

This chapter presents an application of a mathematical programming model to horizontal well placement in shale using a 3D shale dynamic model. The

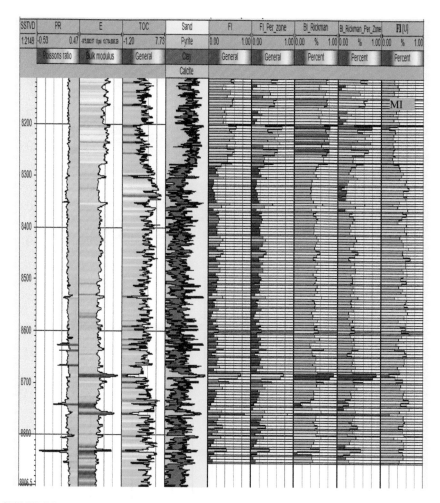

FIGURE 4.8
Property distribution for one vertical well.

new cutoffs contribute to reaching the optimum number of wells passing through sweet-spot portions of shale rock.

- A new mineralogical index map based on mineralogy has been suggested. Such maps are a new technology to guide well placement and fracturing positioning in unconventional resources.
- This work proposes a new quality map to inform decision making for well placement in unconventional resources.
- The new quality map is built from two different schemes for unconventional characteristics.

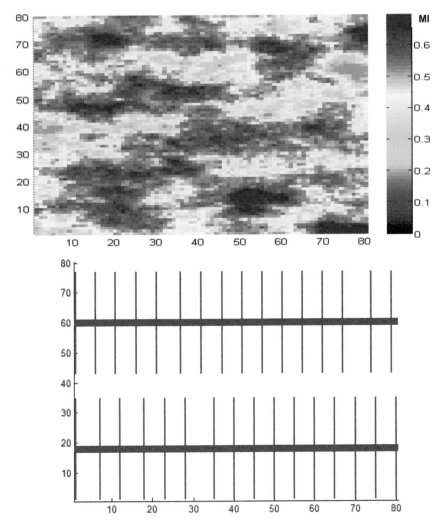

FIGURE 4.9
Recommended well placement in one layer from the shale model to accommodate fracture placement (bottom graph) according to MI map (upper graph).

The current model is still under development. FI is in the early stages, and the measured geomechanical parameters are assumed to be constant here. In order to make the model adaptive, these parameters can be made time dependent to account for dynamic changes within the same model. Since geochemistry is an insufficient guide for sweet spots, further work will integrate geochemical and geomechanical parameters to account for dynamic changes due to depletion.

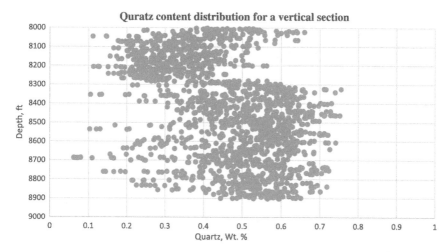

FIGURE 4.10
Quartz distribution for a vertical well in Wolfcamp across Permian Basin.

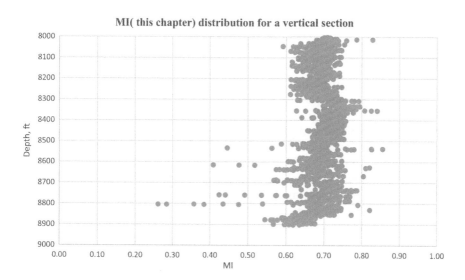

FIGURE 4.11
MI distribution for a vertical well in Wolfcamp of Permian Basin shale.

TABLE 4.5

Recommended Ranges to Be Used in Shale Reservoirs

Brittleness Mineralogical Index, %		Classification	MI (This Chapter)	
70	80	Good rock	0.88	1.00
60	70		0.77	0.88
50	60		0.66	0.77
40	50	Poor rock	0.55	0.66
30	40		0.44	0.55
20	30		0.33	0.44
10	20		0.22	0.33

Acknowledgments

The authors thank Schlumberger for providing the Petrel software used for modeling the shale reservoir.

Appendices

Appendix A: Abbreviations

FI	Fracturability index
E	Young's modulus, psi
E'	Plane strain modulus, psi
E_n'	Normalized plane strain modulus
ν	Poisson's ratio
ρ	Density, lb./ft^3
TOC	Total organic carbon content
MBI	Mineralogical brittleness index
$x_{(j,k)}$	Y_Z location of a fracture in the reservoir represents the location (J, K) in the shale formation grid
X	Coordinate axis along well path, ft.
Y	Coordinate axis along fracture path, ft.
BHP	Bottomhole pressure, psi
Q_g	Gas flow rate, Mscf/d
c_f	Formation compressibility, psi^{-1}
L_f	Fracture half-length, ft.
P_i	Initial reservoir pressure, psi
L	Well lateral length, ft.

$\Delta X, \Delta Y, \Delta Z$	Model grid dimensions in x, y, and z directions, ft.
P_{net}	Net pressure, psi
X_e, Y_e	Rectangular reservoir shape dimensions in x and y directions, ft.
W	Horizontal well
F	Fracture stage
D_{min}	Minimum well spacing, ft.
R_e	Drainage radius
Time	Time of a single run to obtain optimum solution
Max_{Lfrac}	Maximum fracture half-length
$D_{min}f_w$	Minimum distance between a fracture extending from one well and another wellbore
t	Time in seconds for obtaining optimum solution.
N_{wells}	Optimum suggested number of wells
$N_{fractures}$	Optimum suggested number of fracture stages
$N_{fractures}/well$	Optimum suggested number of fracture stages per well
$\emptyset$	Porosity
h	Thickness
K	Permeability
Q	Quartz
Cl	Clay
C	Carbonate
LS	Limestone
MI	Mineralogical index (developed in this chapter)

Appendix B: Formulation of Optimization Problem

The following is the problem formulation.

Problem Definition

Decision Variables:

$x_j^k \in \{0,1\}$, Indicates if cell j is chosen for wellbore k

Sets:

$X^k \subset \{1...N\}$, Set of indices of the cells able to be fractured from wellbore k

$W^k \subset \{1...N\}$, Set of cells containing wellbore k

$M^k \subset \{1...N\}$, Set of cells directly neighboring wellbore k

$E_{11}^j = \{\ell \mid \ell = 1...N, d(j,\ell) < d_{11}\}$, Set of cells that violate the minimum spacing d_{11} with cell j

$E_{12}^j = \{\ell \mid \ell = 1...N, d(j,\ell) < d_{12}\}$, Set of cells that violate the minimum spacing d_{12} with cell j

Construction of set X^k:

$$j \in X^k \text{ if } (1-3) \text{ are all true}:$$

(1) $j \in 1...N$

(2) $d(j, \ell) \leq L_f$ for some $\ell \in W^k$ (cell j is sufficiently close to wellbore k)

(3) $d(j, m) \geq d_{fw}$ $\forall m \in W^p, p = 1...N_w, p \neq k$ (cell j is sufficiently distant from all other wellbores)

Indices:

$j^+(k)$, Index of cell symmetric to cell j across wellbore k

$j^*(k)$, Index of cell connecting cell j to wellbore k along the plane of minimum horizontal stress

Parameters:

$FI_j \in [0,1]$, Fracturability index value for cell j of the reservoir

$C_j^k \in [0,1]$, Cost of fracturing cell j from wellbore k

N, Total number of cells in reservoir model

B, Maximum allowable cost of all fracturing

N_w, Number of wellbores (total wells and laterals)

d_{12}, Minimum spacing between fractures emanating from different wellbores

d_{11}, Minimum spacing between fractures emanating the same wellbore

d_{fw}, Minimum spacing between fractures emanating from one wellbore and all other wellbores

L_f, Maximum fracture half-length

Constraints:

Spacing of chosen node j and other wellbores:

The maximization of the function in Equation 4.2 is subject to constraints in the following equations:

$$x_j^k + x_i^k \leq 1, \quad \forall i \in \left(M^k \cap E_{12}^j\right), \quad j \in M^k, \quad k \in 1,...,N_w \qquad (4.3)$$

Equation 4.4 states that a minimum spacing of fractures emanating from different wellbores is specified:

$$x_j^k + x_i^\ell \leq 1, \quad \forall j \in X^k, \quad \forall i \in \left(X^\ell \cap E_{11}^j\right), \quad \forall k, \ell = 1,...,N_w, k \neq \ell \qquad (4.4)$$

Spacing of fractures emanating from different wellbores is constrained as shown in Equation 4.5 (there should be no two fractures from the same node):

$$x_j^k + x_i^\ell \le 1, \quad \forall i \in \left(X^\ell \cap E_{11}^j\right), \quad j \in X^k, k \in 1,\dots,N_w, \quad \ell = 1,\dots,N_w, \ell \ne k \quad (4.5)$$

Equation 4.6 is shows connectivity of the nodes to the wellbore:

$$x_i^k - x_{i^*}^k \le 0, \quad \forall i \in X^k, \quad \forall k = 1\dots N_w \quad (4.6)$$

Equation 4.7 shows symmetry of the fracture around the wellbore:

$$x_j^k - x_{j^+}^k \le 0, \quad \forall j \in X^k, \quad \forall k = 1\dots N_w \quad (4.7)$$

Equation 4.8 shows total cost of fracturing; it is a cost constraint:

$$\sum_{k=1}^{N_w} \sum_{j \in X^k} C_j^k x_j^k \le B \quad (4.8)$$

Figure 4B.1 shows how the new MI correlates with mineralogical components (Quart/Quartz + Dol + Clay).

Figures 4B.2 through 4B.5 show the relations between the different parameters used for creating our proxy.

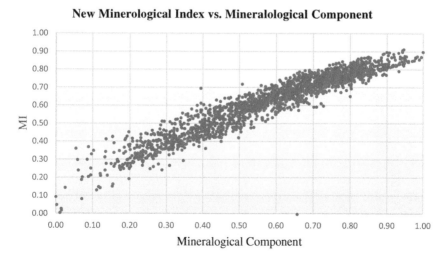

New Minerological Index vs. Mineralological Component

FIGURE 4B.1
Mineralogical index-built correlation.

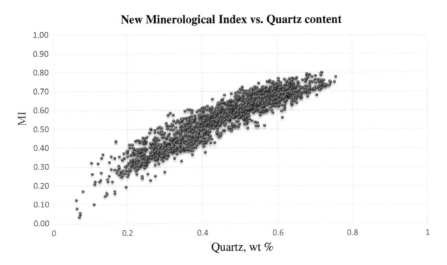

FIGURE 4B.2
MI versus quartz wt. %.

The following figure shows the placement of horizontal wells based on two maps, TOC versus MI. Figures 4B.6 through 4B.11 demonstrate the horizontal well placement entry locations as a result of the optimization approach for the shale model, with a match between the two proxies used for placing the wells for the five cases shown.

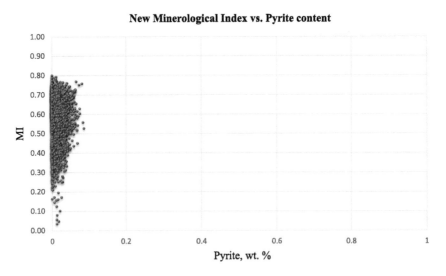

FIGURE 4B.3
MI versus pyrite wt. %.

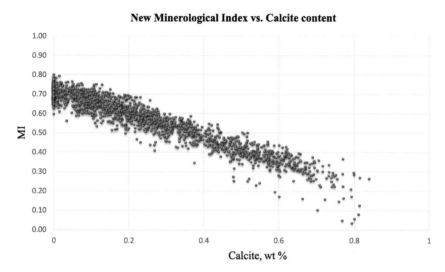

FIGURE 4B.4
MI versus calcite wt. %.

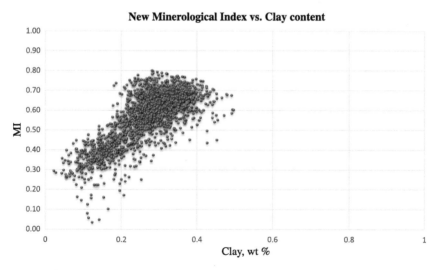

FIGURE 4B.5
MI versus clay wt. %.

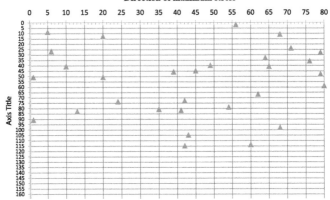

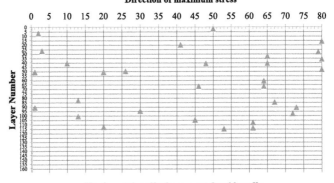

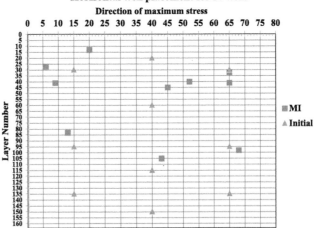

FIGURE 4B.6
Comparison between initial well placement and optimized placement for 10 wells using IP.

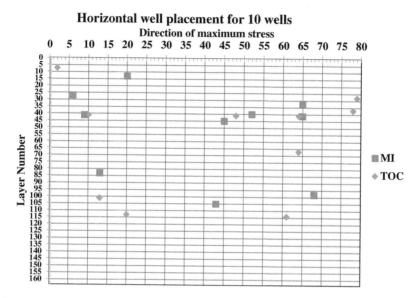

FIGURE 4B.7
Optimum horizontal well placement for 10 wells using TOC and MI as a heat map.

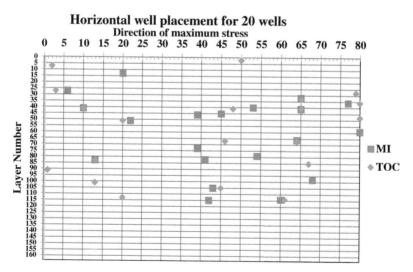

FIGURE 4B.8
Optimum horizontal well placement for 20 wells using TOC and MI as a heat map.

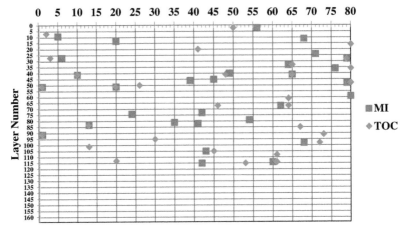

FIGURE 4B.9
Optimum horizontal well placement for 30 wells using TOC and MI as a heat map.

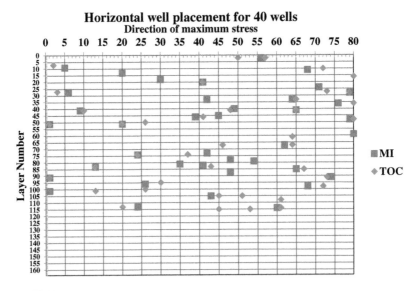

FIGURE 4B.10
Optimum horizontal well placement for 40 wells using TOC and MI as a heat map.

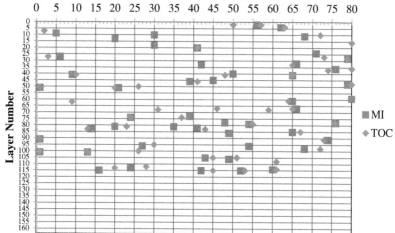

FIGURE 4B.11

Optimum horizontal well placement for 50 wells using TOC and MI as a heat map. Classification of Total Organic Carbon is recommended in Table 4B.1.

TABLE 4B.1

Total Organic Carbon (TOC) Ranges after Next Schlumberger 2013

Based on Early Oil Window Maturity		
Hydrocarbon Generation Potential	TOC in Shale (wt. %)	TOC in Carbonates (wt. %)
Poor	0.0–0.5	0.0–0.2
Fair	0.5–1.0	0.2–0.5
Good	1.0–2.0	0.5–1.0
Very good	2.0–5.0	1.0–2.0
Excellent	>5.0	>2.0

5

Sequencing and Determination of Horizontal Wells and Fractures in Shale Plays: Building a Combined Targeted Treatment Scheme

5.1 Introduction

Bowker (2003) determined that the majority of the production in the Barnett Shale comes from zones with 45% quartz content and only 27% clay. Wang and Gale (2009) and Jarvie et al. (2007) introduced a brittleness index defining ductile and brittle regions in terms of mineralogical analysis. Yu et al. (2013) used economic optimization to identify the optimum parameters of fracture conductivity and distance between two neighboring wells based on Barnett. However, the well-known heterogeneity of shale plays is another very important criterion that has a direct impact on how a shale play should be exploited. The presence of natural fractures, as well as heterogeneity in geomechanical, petrophysical, and geochemical properties, impacts the utilization of shale plays. Stegent et al. (2012) and Cipolla et al. (2012), among others, have hypothesized that at least 50% of created fractures do not contribute significantly to the total production. This observation indicates that the process of uniformly spacing the fractures or even the horizontal wells does not yield the optimum utilization of a shale play. Three issues are therefore necessary to address:

1. Defining a representative shale sweet-spot indicator
2. Location and optimization of the process of locating horizontal wells in a shale play
3. Location and optimization of the process of locating hydraulic fractures in a horizontal well

The third issue of locating fractures has been discussed by several authors. Rickman et al. (2008) defined a new brittleness based solely on Young's modulus and Poisson's ratio. By use of this brittleness index, various areas in a shale play may be divided into brittle and ductile areas. Fractures would be recommended in the areas defined as brittle (Figure 5.1) shows the graph by

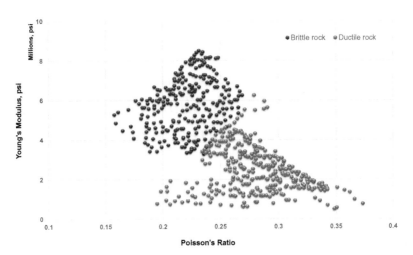

FIGURE 5.1
Brittleness index. (Modified from Rickman, R. et al. 2008. A Practical Use of Shale Petrophysics for Stimulation Design Optimization: All Shale Plays Are Not Clones of the Barnett Shale. *Presented at SPE Annual Technical Conference and Exhibition (ATCE)*, Denver, CO, September 21–24, doi:10.2118/115258-MS.)

Rickman et al. indicating the distribution of brittle and ductile areas of the shale plays in areas they examined.

Recently, Mullen et al. (2012) introduced a more complex fracturability index. The primary rock property input in their work is brittleness as directly correlated with Brinell hardness. Mullen et al. determined that zones where proppant embedment is less likely to occur and the rock is more likely to create a complex fracture network have a high fracturability index value.

Parker et al. (2009) used well logging to show the variation in Young's modulus and Poisson's ratio along a Haynesville shale horizontal (Figure 5.2).

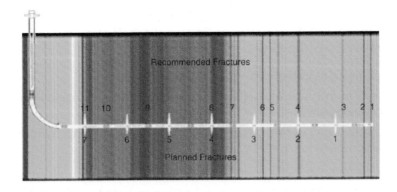

FIGURE 5.2
The brittleness track shows variation in rock heterogeneity; the red section is more brittle. (After Parker, M., Petre, J.E., Dreher, D. 2009. Haynesville Shale: Hydraulic Fracture Stimulation Approach. *Presented at the International Coalbed & Shale Gas Symposium*, Tuscaloosa, AL.)

Parker et al. used the criteria developed by Rickman et al. and Mullen et al. to locate the relatively brittle areas along the horizontal. The brittle areas were selected for fracturing. See Figure 5.2 for planned versus optimum fractures.

Numerous authors have developed various indices to define areas of fracturability. While helpful in approximating future locations of planned fractures, such approaches tend to focus on one area of technology. None of those techniques consider all the possible parameters that may affect fracture placement and propagation.

Optimization methodologies are being developed for fast-tracking the number of fractures and effective horizontal wells in shale plays by a fast and effective method. The sequencing of well and fracture placement is key to achieving economical utilization of shale assets in North America. Of critical importance are the computational methods used for optimization of placement.

While equally important, the second issue, locating horizontal wells in unconventional reservoirs to take advantage of sweet spots, has not received appropriate attention. Serious consideration of this issue has been presented by Alzahabi et al. (2014). In their study, Alzahabi et al. considered the petrophysical, geomechanical, geochemical, and physical properties of the rock in the screening and placement of the fracture and developed a candidate evaluation algorithm.

The designed algorithm explores the geomechanical, petrophysical, and geochemical parameters of a newly discovered shale resource versus major productive shale plays in North America, noting any potential for newly discovered shale plays. Their work is based on a built-in shale success factor, which depends on statistics, database structure, a candidate evaluation approach, and the developed algorithm. The data structure of their algorithm consists of shale play spider plots, completion strategies, mineralogy comparisons, mechanical properties, characteristics, shale gas production indicators, and sweet-spot identifiers. The algorithm design contains characteristics and strategies to identify similarities to prioritized shale parameters and predict the likelihood of similarity, then recommends future development approaches. Clusters of checking maturity parameters, similarity, and then rank and guide are the main root model parameters. A procedure to redefine the groupings to create clusters that better represent the data is a new insight of this work, in addition to prediction of new basin performance based on the performance of proven basins. This procedure will help identification of sweet spots in new basins and guide fracturing and well placement in shale plays.

Due to variations in mechanical, geochemical, and geomechanical properties that occur on the level of a few inches in the shale model, the planning of clusters should not involve their common even distribution. The effect on production is significant when the entire process is optimized.

Figure 5.3 shows a common design in horizontal wells that can be implemented using the perf-and-plug approach. This design accommodates as many fractures as possible to achieve maximum contact surface. At this point, it is pertinent to address envisioned methods of selectively fracturing and plugging.

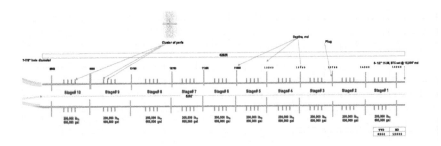

FIGURE 5.3
Commonly used approach of placing fractures in horizontal wells.

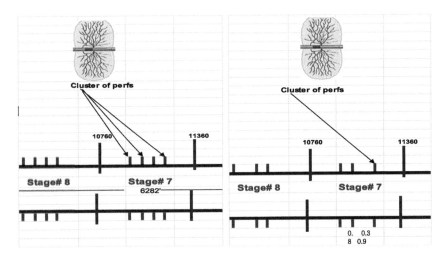

FIGURE 5.4
Comparison between equally spaced clusters within the stage. At the right are shown FI drove-placed clusters.

Figure 5.4 shows a comparison between equally spaced clusters within the stage. At the right are shown FI drove-placed clusters. The math demonstrates that the conventional approach is preferable, but accurate basin modeling tells a new and different story guided by the newly developed fracturability index approach, applicable with brittle versus ductile rock, naturally fractured rock versus none, and very heterogeneous versus uniform properties.

5.2 The Developed Approach

Achieving the maximum contact area with high-quality points within the model is possible, where optimization of vertical and deviated well placement helps achieve better access to good-quality points. The quality of the shale resource comes from brittle and high TOC content. The key is to allow representative factors to guide optimization algorithms. Adapting the resource model can benefit placement and guide fracture positioning in any shale play.

The technology takes into account various elements such as size, time, and number. In contrast to current thought in the industry, which posits that more fracture is better, this work develops a technology in support of the notion that the effective fracture is better. Effective fractures are those that target brittle zones in the shale basin, resulting in branched fractures and deep encroachment into virgin parts of the model.

5.2.1 Data Analysis

This study researched the effects of multiple parameters such as Poisson's ratio and Young's modulus on fracturability index.

Insights of Figures 5.5 and 5.6:

- Both E and PR affect FI
- There is a relationship between FI and E within a certain range

As a result of new ways of understanding the rock, fluid, and interactions, our ability to design treatment will not be limited. Standardizing the treatment process with newly developed indices facilitates the initiation of fracture in shale according to a planned, automated process.

An integrated mix of high-resolution seismic, accurate sonic, and formation imaging is the main enabling tool for detailed basin description. Structural features can be translated into quantified ranges, which should be filtered

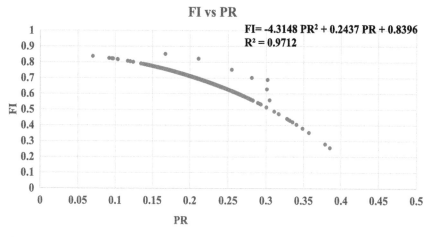

FI vs PR

$FI= -4.3148 PR^2 + 0.2437 PR + 0.8396$
$R^2 = 0.9712$

FIGURE 5.5
Effect of Poisson's ratio on FI.

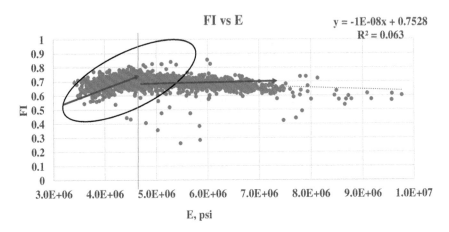

FIGURE 5.6
Effect of Young's modulus on FI.

within the algorithm to subcategories. The classifications include *very good brittle, brittle, ductile,* and *very good ductile.*

5.2.2 Developing an Integrated Fracturability Index Correlation

The main parameters of rock are as follows:

Independent Variables: Plane strain Young's modulus, mineralogical brittleness index

Dependent Variables: Fracturability index = Normalized strain energy release

The term G_c is the critical energy release rate; it is related to U_a, or energy per fracture area.

$$G_c = U_a \tag{5.1}$$

$$U_a = \frac{K_{Ic}^2}{4\mu(1+v)} \tag{5.2}$$

Equations 5.1 and 5.2 after Wickham et al. (2013).
The energy released in creating a fracture is approximately equal to the energy required to create it.

5.2.2.1 First Correlation

$$FI = G_{c_n} = \frac{G_c - G_{cmin}}{G_{cmax} - G_{cmin}} \tag{5.3}$$

$$G_c = \frac{K_{IC}^2}{E} \tag{5.4}$$

$$FI = 10^{\left(a+b\times\log((1/E)/1-v^2)+c\times\log\left(\frac{quartz+Pyrite}{quartz+Pyrite+Clay+Calcite}\right)\right)} \tag{5.5}$$

	Coefficients
A, intercept	−0.2332
B	−0.0324
C	1.0098509
Regression statistics	
Multiple R	0.98

The first correlation is built on the concept of the critical energy to create a fracture: as an example, fracturability index ≥ 0.41 indicates higher initial energy required, which stands for brittle rock. On the other hand, $FI \leq 0.41$ indicates less initial energy required, which stands for ductile rock.

5.2.2.2 Second Correlation

The energy release rate failure criterion states that a crack will grow when the available energy release rate exceeds the critical energy release rate value.

$$G = \frac{\partial(U-V)}{\partial A} \tag{5.6}$$

where U is the potential energy available for crack growth, V is the work associated with any external forces acting in the crack area (crack length for 2D problems), and the G unit is J/m².

$$G \geq G_c \tag{5.7}$$

G_c is the critical value and fracture energy:

$$G = \frac{K_I^2}{E'} \quad \text{(Assume plane strain)} \tag{5.8}$$

$$K_I = 0.313 + 0.027 \times E \quad \text{Correlation} \tag{5.9}$$

Then, substitute Equation 5.9 in Equation 5.8:

$$G = \frac{K_I = [0.313 + 0.027 \times E]^2}{E'} \sim E \tag{5.10}$$

$$G \propto E \tag{5.11}$$

$$FI = B_0 \times 10^{[B_1 + B_2 \times \log(1/MFI) + B_3 \times \log(E')]} \tag{5.12}$$

where
  E is in psi
  MI = Mineralogical index

Higher E~, but the better the resultant fracture, the more energy required for growth.

$$G \propto E$$

The more brittle quartz is in composition and the easier it is to fracture, the less energy is required

$$G \propto \frac{1}{MBI} = \frac{Quartz + Feldspar + Pyrite + Dolmoite + Calcite}{Quartz + Feldspar + Pyrite} \tag{5.13}$$

Consider the following:

- The energy that is consumed by creating a new surface should be balanced by the potential energy of the system.
- $dW_{elas} + dW_{ext} + dW_s + dW_{kin} = 0$.
- dW_{elas} represents the change in elastic energy stored in the rock.
- dW_{ext} is the change in potential energy of exterior forces.
- dW_s is the energy dissipated during the propagation of a crack.
- dW_{kin} is the change in kinetic energy.
- Applying the Griffith criterion (Griffith 1921) for fracture initiation and growth:

B_0	10^{-6}
B_1	-0.233205322
B_2	-0.032424691
B_3	1.009850946

$$FI = B_0 \times 10^{[B_1 + B_2 \times \log(1/MFI) + B_3 \times \log(E')]} \tag{5.14}$$

The second correlation is built on the concept of the energy needed to create a fracture: as an example, FI = 7 indicates higher energy required, which stands for ductility. On the other hand, FI = 2 indicates less energy required, which stands for brittle rock.

5.2.3 An Alternative Industry-Used Approach for Locating Sweet Spots

In this section, a new integrated approach for understanding unconventional rock is established, including geomechanical parameters and detailed mineralogy of the reservoir.

The new integrated FI takes both geomechanical and geochemical effects into consideration.

Three categories are distinguished: *highest, high zones,* and *bad zones.*

The new classification identifies the shale reservoir based on brittleness, high porosity, and organic material. The new correlation makes use of the results of a study by Walls et al. (2012). With their own developed criteria as a separate scale for checking sweet spots according to developed FI in Permian Basin shale reservoirs, they used 16 core samples from Eagle Ford, located in Maverick, Dimmit, LaSalle, and Atascosa counties in South Texas. Pore- and grain-scale results were used to obtain information on porosity distribution, organic material volume, and organic material pore structure.

For shale reservoirs, *sweet spots* refers to those portions of the basin that have high porosity and highly brittle composition and are high in organic matter and easier to fracture. The higher the porosity, the greater the maturity.

Density and photoelectric effect from dual-energy CT allowed independent verification of our data for sweet-spot portions. Then, mineralogical and geomechanical principles were used to construct the new correlation. The highest category contains high porosity, high kerogen, and a composition of quartz; the second in rank contains high porosity, high kerogen, and a composition of calcite, after Lewis (2013) and Schlumberger (Table 5B.1; see Appendix B).

1. Highest category sweet-spot zones have bulk density (*RHOb*) < 2.53 and photoelectric index (PE) < 4.0.
2. High category sweet-spot zones have bulk density *RHOb* < 2.53 and PE > 4.0.
3. Low category sweet-spot zones: the remaining data.

5.2.4 Photoelectric Index for Mineral Identification

The photoelectric index (PE) is a measurement by modern logging tools to measure the absorption of low-energy gamma rays by the formation in units of barns per electron. Significant here is the difference between main reservoir rock-forming mineral quartz (PE = 1.81 barns electron^{-1}), calcite (PE = 5.08 barns electron^{-1}), dolomite (PE = 3.14 barns electron^{-1}), montmorillonite (PE = 2.04 barns electron^{-1}), kaolinite (PE = 1.49 barns electron^{-1}), illite (PE = 3.45 barns electron^{-1}), chlorite (PE = 6.3 barns electron^{-1}), and oil (PE = 0.12 barns electron^{-1}) (for more details, see *Physical Properties of Rocks: A Workbook* by J.H. Schön 2011).

5.2.5 Combining Both Techniques for Sweet-Spot Identification

Table 5B.2 (see Appendix B) introduces the recommendation list of shale sweet-spot classification according to the developed index (this chapter) versus that of Walls et al.

Figure 5.7 explains the developed sweet-spot selection methodology. Three new classification criteria, S1, S2, and S3, are introduced and tested against commonly used industry practice.

Figure 5.8 demonstrates a geomechanical-based FI correlation introduced by Alzahabi et al. (2015b).

Figure 5.9 demonstrates the relationship between the second developed fracturability index versus plane strain Young's modulus.

Figure 5.10 demonstrates the relationship between the second developed fracturability index versus mineralogical brittleness index.

Figure 5.11 shows the preferred facies, which is organic siliceous shale and organic mixed shale, as introduced by Wang and Carr (2013). Figure 5.11 also shows shale lithofacies after Wang and Carr (2013).

Figure 5.12 shows a classification of four categories of sweet-spot regions after Walls et al. (2012). The red and green indicate recommended sweet-spot locations of shale.

Figure 5.13 shows shale lithofacies classification after Wang and Carr (2013).

The scientific reason for the use of density and PE in this chapter as independent variables is that each works as a direct indicator of formation lithology.

Figure 5.14 shows a geographic map of the studied area.

Figures 5.15 and 5.16 show testing of the data according to our three devolved criteria versus the criteria developed by Walls et al. (2012).

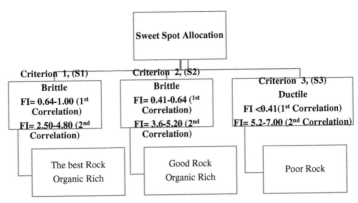

FIGURE 5.7
The developed sweet-spot selection methodology for placing wells and fractures in shale reservoirs (this chapter).

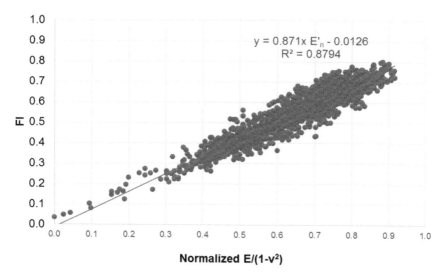

FIGURE 5.8
A geomechanical-based FI correlation introduced by Alzahabi et al. (2015b).

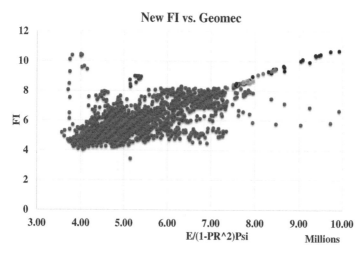

FIGURE 5.9
FI (second correlation) versus plane strain Young's modulus.

5.3 Study Area

The area of study is located in the Midland area, Permian Basin shale Wolfcamp formation. Wolfcamp is dominated by organic-rich and siliceous lithofacies. Its detailed description is included in Table 5B.3 (see Appendix B) and Figure 5.14.

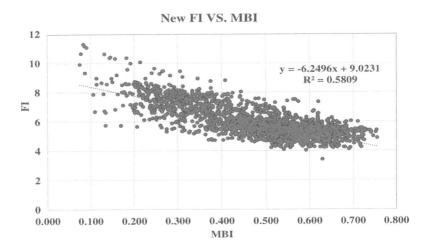

FIGURE 5.10
FI (second correlation) versus mineralogical brittleness index.

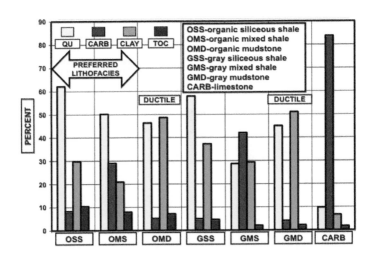

FIGURE 5.11
Marcellus shale lithofacies. (After Wang, G., Carr, T.R. 2013, Prediction and Distribution Analysis of Marcellus Shale Productive Facies in the Appalachian Basin, USA, Presentation at AAPG Annual Convention and Exhibition, Pittsburgh, Pennsylvania, May 19–22.)

5.3.1 Testing and Validation of the Work (Permian Basin Wolfcamp Shale Reservoir Data)

Table 5B.4 (see Appendix B) lists the testing process of the two criteria used in classifying the shale reservoirs. The next section offers a detailed explanation of how this developed integrated FI can be used to guide fracture and wells in a sequence.

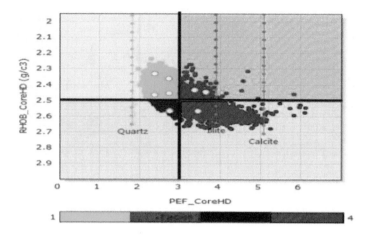

FIGURE 5.12
Four categories of sweet-spot regions. (After Walls, J.D., Sinclair, S.W., Devito, J. 2012. Reservoir Characterization in the Eagle Ford Shale Using Digital Rock Methods. *WTGS2012 Fall Symposium.*)

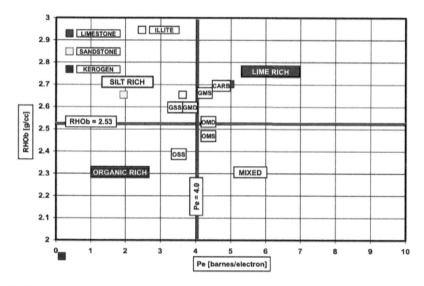

FIGURE 5.13
Shale lithofacies. (After Wang, G., Carr, T.R. 2013, Prediction and Distribution Analysis of Marcellus Shale Productive Facies in the Appalachian Basin, USA, Presentation at AAPG Annual Convention and Exhibition, Pittsburgh, Pennsylvania, May 19–22.)

Figures 5.15 and 5.16 show application of the criteria on the tested model.

Figures 5.17 through 5.21 demonstrate detailed application of the developed approach, starting with S1, S2, and S3, matching all criteria to suggest the best horizontal well path. The result of this testing indicates that it is preferable to locate wells in the sweet spots of the reservoir, where quartz is abundant,

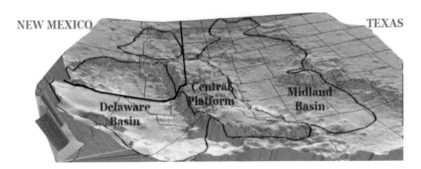

FIGURE 5.14
Map of a location of Permian Basin reservoir.

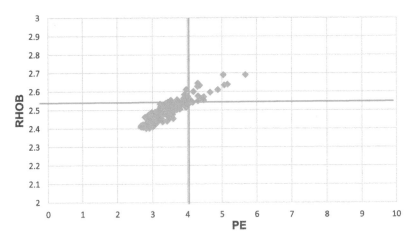

FIGURE 5.15
Selected zone criteria 1.

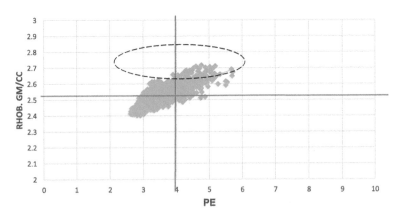

FIGURE 5.16
Shale formation criteria 1, 2, and 3.

FI distribution for a vertical section

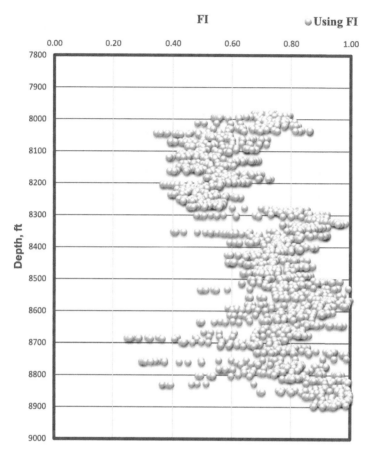

FIGURE 5.17
Newly developed FI versus depth.

which facilitates the drilling of horizontal wells and the locating of fractures, and where hydrocarbons are concentrated. It is also supported that sweet-spot portions of rock occur where horizontal stress is minimum and the differential horizontal stress ratio is maximum, as explained later in this chapter.

5.4 Differential Horizontal Stress Ratio

5.4.1 Net Pressure and Stress

By the use of net pressure data, it is possible to gain an idea of the difference between maximum and minimum stress.

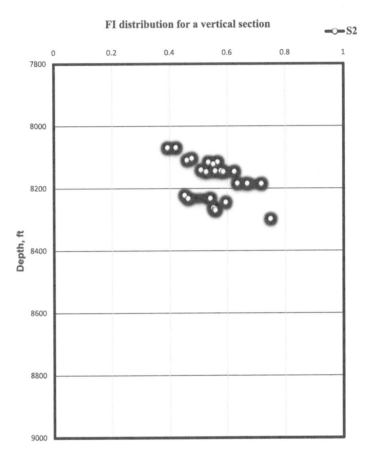

FIGURE 5.18
Newly developed FI screened by S1 conditions versus depth.

Nolte Smith provided the simple Equation 5.15 for fissure opening:

$$P_o = \frac{\Delta \sigma_h}{(1 - 2\nu)} \tag{5.15}$$

The differential horizontal stress ratio, a scale used for placing horizontal wells, is crucial in deciding whether a selectively chosen region in a shale play will fracture easily. DHSR can be obtained from seismic data alone.

A planned fracture is believed to exist when designing fracks in high-quality relative points of DHSR. The following formula has been used here to calculate DHSR.

$$DHSR = \frac{\text{Maximum stress} - \text{Minimum stress}}{\text{Maximum stress}} \tag{5.16}$$

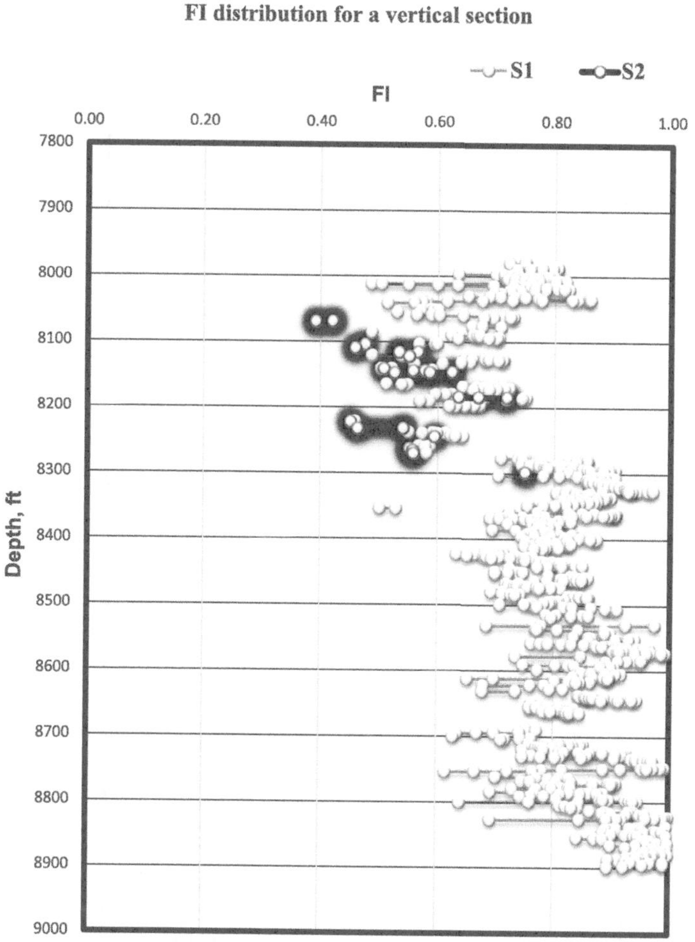

FIGURE 5.19
Newly developed FI screened by S1 and S2 conditions versus depth.

Figure 5.22 illustrates four methods to locate the lateral. The log tracks show that the horizontal well should laced at 8350 ft.

Figure 5.23 illustrates the differential horizontal stress profiles. This plot suggests placing the horizontal well in the interval 8300–8400 ft.

5.5 Hydraulic Fracturing Stage Sequencing

With many wells (ranging from 5 to 40) originating from one pad as a surface location, a common successful scenario involves *simul frack,* which is the

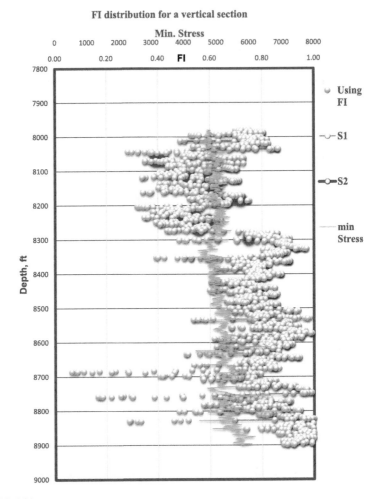

FIGURE 5.20
Newly developed FI screened by S1, S2, and minimum stress profile versus depth.

fracturing of stages simultaneously on parallel wells; the second approach involves alternating stages between adjacent parallel wells, or *zipper frack*; the third (recently developed) is *modified zipper frack*. These approaches are successful in utilizing the surface equipment and developing a larger surface area in shale and tight formations. The key behind these approaches is that the rock between wells may have a superposition of stress, which can help in effective stimulation of tight rocks. Many microseismic results show that the second and third stages are less effective than the first treatment even if the same fluid volume and proppant amount are injected.

This chapter introduces a new fracturing methodology called cascade fracturing technology. This methodology starts by dividing the well path into segments, identifying the order of fracture locations along the well

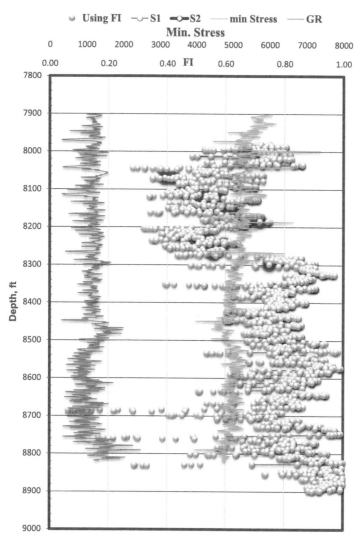

FIGURE 5.21
Newly developed FI screened by S1, S2, minimum stress and gamma ray profile versus depth.

path, and ordering the fractures from the production point of view. This methodology is followed for all wells in the reservoir, thus prioritizing the completion strategy. Results show that shale reservoirs may be produced more effectively by the use of this new methodology. In addition, the number of fracture stages designed by use of the new fracturability index is lower than by use of conventional techniques, thus reducing the cost of fracturing. The timing of fractures, number of fracture stages, clusters for each stage, and number of wells may be determined based on reservoir

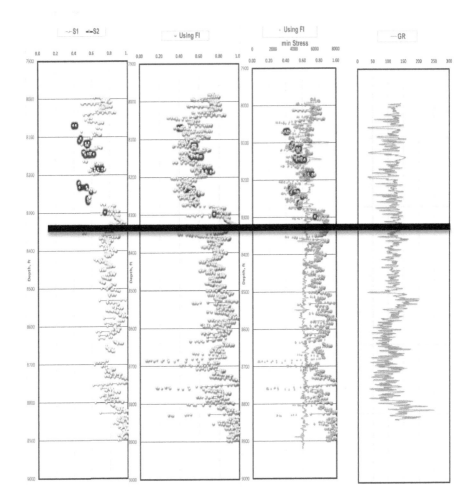

FIGURE 5.22
Permian Wolfcamp shale: Midland Basin, Texas.

and fluid properties rather than by trial-and-error approaches. Figure 5.24 shows a schematic that follows the suggested design after sequencing by the newly developed FI.

Figure 5.25 shows a map of the FI calculated for the tested shale model using correlation by Alzahabi et al. (2015b).

Figure 5.26 shows an optimized placement of wells and fractures in the same model.

Figure 5.27 shows the rank of fracture developed by this work, where FI values suggest the cascade fracturing approach through order of fracture placement stages from toe to heel in one direction (fractures 4 and 5 are not operationally recommended). The numbers rank the possibility of placing fracture stages according to their potential and not on fracture creation time.

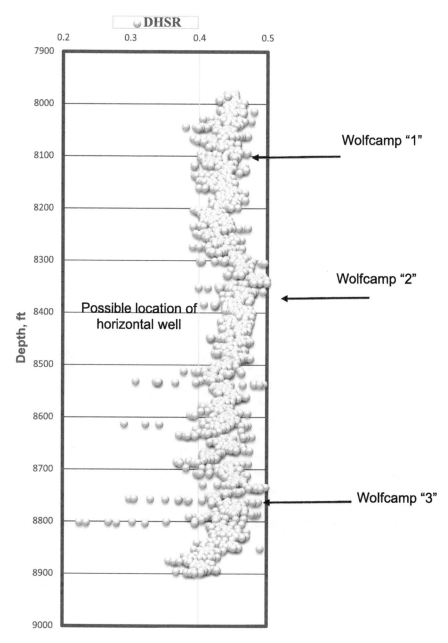

FIGURE 5.23
Differential horizontal stress ratio versus depth.

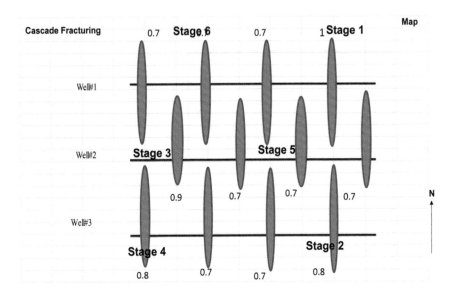

FIGURE 5.24
Schematic of a map view of suggested fracture stages suggested by FI values.

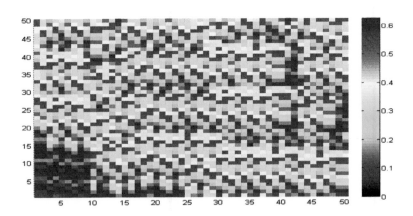

FIGURE 5.25
Fracturability index map based on the tested shale model.

5.6 Conclusions

The developed work is a new sweet-spot guide, designed primarily for placement of deviated wells and fracture stages in shale resources. Issues such as height growth and variations in rock mechanics have been extensively checked here.

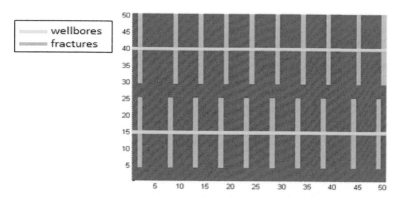

FIGURE 5.26
Optimal placement for a maximum fracture half-length of 10 cells and no overlap.

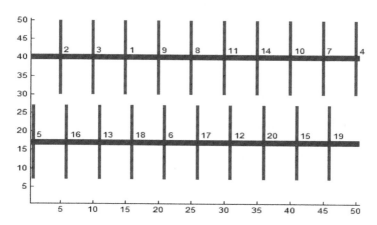

FIGURE 5.27
Sequencing of fractures according to FI values.

In conclusion, the issues of well and fracture placement and optimization require more focused attention than they currently receive. Increased attention to this area of investigation should lead to more economical development of shale plays.

Currently, it is well known that fracture gradient and in-situ stress along the path are not the same. We therefore recommend that each single stage should follow a certain method of fracturing, and every single cluster of perf should be consistent.

Two new screening criteria are being developed for identifying sweet-spot locations within shale reservoirs.

An agreement has been obtained for locating the horizontal wells between FI, S1, S2, and DHSR, a promising indication that the new criteria can be used for future well placement and fracture allocation.

Appendices

Appendix A: Abbreviations

FI	Fracturability index
E	Young's modulus, psi
E'	Plane strain modulus, psi
E'_n	Normalized plane strain modulus
ν	Poisson's ratio
ρ	Density, lb./ft^3
$X_{(i,j)}$	X_Y location of a fracture in the reservoir represents the location (I, j) in the shale formation grid
X	Coordinate axis along well path, ft.
Y	Coordinate axis along fracture path, ft.
BHP	Bottomhole pressure, psi
Q_g	Gas flow rate, Mscf/d
c_f	Formation compressibility, psi^{-1}
L_f	Fracture half-length, ft.
P_i	Initial reservoir pressure, psi
L	Well lateral length, ft.
$\Delta X, \Delta Y$	Model grid dimensions in x and y directions, ft.
P_{net}	Net pressure, psi
X_e, Y_e	Rectangular reservoir shape dimensions in x and y directions, ft.
W	Horizontal well
F	Fracture stage
D_{min}	Minimum well spacing, ft.
G_c	Critical energy release rate
U_a	Energy per created fracture area
$\varnothing$	Porosity

Appendix B: Classifications of Sweet Spots

TABLE 5B.1

An Example of Classified Sweet Spots by Walls et al. (2012)

High porosity	and/or kerogen	More quartz
Lower porosity	or kerogen	More calcite
Lower porosity	or kerogen	More calcite
Lower porosity	or kerogen	Less calcite

TABLE 5B.2

Recommendation for Sweet-Spot Identifiers and Shale Classification

Category	Classification	FI, First Correlation	FI, Second Correlation	P_e and ρ_b
Highest category sweet-spot zones	S1	0.64–1.00	2.50–4.80	$\rho_b < 2.53$ and $P_e < 4.0$
High category sweet-spot zones	S2	0.41–0.64	3.60–5.20	$\rho_b < 2.53$ and $P_e > 4.0$
Low category sweet-spot zones	S3	0–0.41	5.20–7.00	The rest of the data

TABLE 5B.3

Data Ranges Used for Testing the New Criteria

Input Parameters	
TOC, wt.%	2.3
RO	1.3
Total porosity, %	10
Net thickness, ft.	1400
Adsorbed gas	N/A
Gas content	N/A
Depth, ft.	10,100–11,500 ≈ 10,800
E, PR	4.6 E + 6, 0.238
Mineralogy	Quartz, feldspar, 61%
	Clay, 30%
	Pyrite, 1%
	Carbonate, 6.28%
	Kerogen, %

TABLE 5B.4

Checking for Sweet Spots Using Two Scales

First Criteria, $Rhoz < 2.53$ g/cc and $P_e > 4$	Second Criteria $Rhoz < 2.53$ g/cc and $P_e < 4$
Yes	No
Yes	No
Yes	No
No	No
No	No
No	No
No	No
No	No
No	No
No	No

6

A Computational Comparison between
Optimization Techniques for Well
Placement Problem: Mathematical
Formulations, Genetic Algorithms,
and Very Fast Simulated Annealing

6.1 Introduction

After characterization of a reservoir through seismic, radiological, and other means of survey, the drainage of the reservoir must be carefully planned. During this process, factors that are considered include the connectivity of reservoir volumes, the overall drainage volume that could be achieved from each potential well location, and constraints due to the presence of other fluid volumes (Wang 2002). Often, in order to simplify this process, each possible well location is screened for suitability and production capability, then ranked accordingly, in a process referred to as *sweet-spot identification*. Quality factor maps were introduced by Gutteridge and Gawith in order to rank well locations based on connectivity rating and productivity index (Gutteridge and Gawith 1996). Alternatively, given 3D models of lithofacies, porosity, and permeability, Deutsch proposed the geo-objects method in which cells in the computational grid model of the reservoir were combined into separate bodies based on their measure of connectivity (Deutsch 1998). In a more recent work, Henery et al. integrated geological model and reservoir properties based on log data to build 3D reservoir property volumes considering facies, then added monthly production to identify sweet spots, a method that works for well optimum locations (Henery et al. 2011).

Once a value is assigned to each potential well location, the optimal combination of well locations that are to be drilled and produced must be determined. This is the well placement problem. Most often, it is assumed that the interaction between separate wells can be ignored, as this assumption greatly reduces computational complexity (Wang 2002). Vasantharajan and Cullick proposed the combination of quality maps with an integer programming solution in order to solve the WPP (Vasantharajan and Cullick 1997). For computationally complex problems where conventional optimization techniques require large solution times, metaheuristics techniques offer a way to potentially reach an acceptable solution more efficiently (Talbi 2009). Of this category, genetic algorithms and simulated annealing (SA)-type techniques are two common methods for solving a wide variety of optimization problems (Sen and Stoffa 1995). Genetic algorithms have been shown to outperform human engineers by achieving a higher net present value and oil recovery index (Emerick et al. 2009) and are reported as the most commonly used method for well placement optimization (Alqahtani 2012). GAs have been shown to perform similarly to a covariance matrix adaptation solution, which is another type of evolutionary algorithm (Ding 2008). Earlier, Bangerth et al. reported better results were achieved from the use of two types of simulated annealing algorithms (Bangerth et al. 2006).

An earlier work compared the use of genetic algorithms and integer programming to solve this problem (AlQahtani et al. 2012). For each case studied, it was found that IP was capable of finding solutions at least as good as those obtained from genetic algorithms. The current work introduces a simulated annealing algorithm to the comparison, while also using a redesigned genetic algorithm. The next section contains the formulation of the optimization problem as well as details on the simulated annealing, genetic algorithm, and integer programming techniques used to solve the well placement problem. After that, results of a case study are presented to provide comparison of the efficacies of the three algorithms. Finally, a short summary and conclusions are presented.

6.2 Algorithm Design

The genetic algorithm and simulated annealing-type algorithms both make random changes to the set of chosen well locations. Performance of either algorithm depends heavily on the degree of effort devoted to tuning its parameters to a particular problem. The problem formulation features constraints that become difficult to satisfy when either the minimum well spacing or the size of the chosen set becomes large, especially when changes to the chosen set are made randomly. The difficulty of checking the distance constraints presents a heavy computational burden.

6.2.1 Representing Well Locations

Two approaches were considered for representing the set of potential well locations. One approach enumerates the potential x and y coordinates of potential well locations $i_x \in 1...N_x$ and $i_y \in 1...N_y$, where i_x is the index of the potential well location's x-coordinate, i_y is the index of the potential well location's y-location, and N_x and N_y are the total number of potential well x-coordinates and y-coordinates, respectively. Then, the set of selected locations $\{l_k\}_{k=1}^{Nwell}$ is a series of pairs $\left\{\left(i_x, i_y\right)_k\right\}_{k=1}^{Nwell}$, where integer k indexes the selected wells. In this case, performing mutations in genetic-type algorithms or perturbing the system parameter in simulated annealing-type algorithms corresponds to changes in integers i_x or i_y for one or more selected well locations k.

An alternative method for representing potential well locations is to index every possible well location by integer m, and for the case where potential well locations are aligned on a square grid, m ranges from 1 to $(N_x \times N_y)$. Now, the selected set of wells is a set of integers $\{m_k\}_{k=1}^{Nwell}$. This second method was chosen for use within the genetic and simulated annealing algorithms applied for this work. Every potential well location still has a pair of associated x- and y-coordinates, $(i_x, i_y)_m = (i_{x,m}, i_{y,m})$. However, indexing the potential well locations with a single index m allows for a new way to program the check of minimum spacing constraints. In this work, the satisfaction of the spacing constraints between every pair of potential wells m and n ($m \in 1...(N_x \times N_y), n \in 1...(N_x \times N_y)$) is precomputed and the result stored in coexistence matrix C, such that element $C_{m,n} = C_{n,m} = \begin{cases} 1, d(m,n) \geq d_{min} \\ 0, d(m,n) < d_{min} \end{cases}$,

where $d(m, n)$ is the distance between potential well locations m and n. Then, during the GA or SA algorithms, checking if the minimum distance constraint is met between potential locations m and n is reduced to seeing whether $C_{m,n}$ is 1. This step eliminates the need to repeatedly calculate the distances between the same wells.

Along with formation of the coexistence matrix, both the genetic and SA algorithms are initiated with the creation of an initial set of selected wells, referred to as an individual in the GA and the parameter in SA. This initial set is formed by a greedy method, as outlined in Algorithm 6.1. While the SA algorithm requires only the use of three parameters at any one time, the GA requires the creation of an entire population of individuals. In the current application, this population consists of different possible sets of well locations.

Algorithm 6.1 Pseudocode for Greedy Creation of Selected Well Location Set.

$S = $ **GREEDY_SET** (d_{min}, N_{well})

1   enumerate well locations $1 ... N_x \times N_y$ based on descending values of f_{well}
    % create coexistence matrix C:

2   **for** $m = 1 ... N_x \times N_y$

3   **for** $n = m ... N_x \times N_y$

4          $C_{m,n} = C_{n,m} = \begin{cases} 1, & d(m,n) \geq d_{min} \\ 0, & d(m,n) < d_{min} \end{cases}$

5   place location 1 in the set of selected well locations S

6   **for** $m = 2 ... N_x \times N_y$

7   **if** $C_{m,n} = 1$ for all $n \in S$

8   add m to set S

9   **if** size $(S) = N_{well}$

10 | break; % creation of set S is complete

6.2.2 Genetic Algorithm Design

Previous work by the authors that applied a genetic algorithm included modules for mutation, crossover, local search, population intrusion, and selection (AlQahtani et al. 2012). Of these types of operations, the genetic algorithm applied in the current work used only the mutation and selection modules, given its focus on achieving high performance from the mutation operation. The mutation operation on a given set S consists of adding a new well location m and removing any wells of set S that are within the minimum spacing of m. This step is referred to as the well replacement operation, the algorithm of which is detailed in Figure 6.1.

To create an initial population of individuals for the genetic algorithm, the initial individual is mutated in parallel Npop (number of populations)/6 times. For each mutation, a well location that is not in the set of selected locations is added to this set. Once the initial population is created, the genetic algorithm loops over the number of generations. During each generation, a given set of

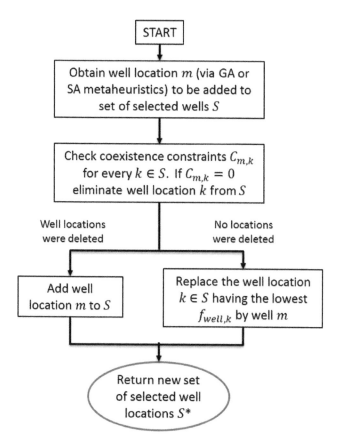

FIGURE 6.1
Algorithm flowchart for the well replacement operation.

well locations is mutated through the well replacement operation, five times serially. This process increases the population size from Npop/6 to Npop. Due to the spacing constraints, a given set S ends up with fewer than Nwell (number of wells in the reservoir) well locations after the well replacement operation. Therefore, after well replacement, an attempt is made to fill each set using a greedy method, wherein the highest objective-valued well locations that meet the constraints with all of the existing locations within the set are added to the set. At the end of every generation, a selection of the sets that move on to the next generation is made. The highest Npop/6 individuals are chosen for the start of the next iteration, and the remaining individuals are discarded. After the algorithm runs for the prescribed number of generations or exceeds the maximum time, the best set of well locations is chosen based upon the highest objective value of all existing individuals. The pseudocode for the genetic algorithm is presented as Algorithm 6.2.

Algorithm 6.2 Pseudocode for Genetic Algorithm.

$S^* = $ GA (d_{min}, N_{well})

1 $S_1 = $ GREEDY_SET (d_{min}, N_{well}) with obj. value F_1
   % create initial population of individuals:

2 **for** $j = 2 \ldots Npop/6$

3 | randomly choose well location $m \notin S_1$

4 | $S_j = $ WELL_REPLACEMENT (S_1, m)

5 | fill open locations in set S_j with greedy
      | search

6 **for** $i = 1 \ldots$ Number_of_generations

7 | **for** $j = 1 \ldots Npop/6$
    | % MUTATION:

8 | **for** $k = 1 \ldots 5$

9 | | $\ell = (k-1) \times {}^{Npop}/_6 + j; \ell' = k \times {}^{Npop}/_6 + j$

10 | | randomly choose well location $m \notin S_\ell$

11 | | $S_{\ell'} = $ WELL_REPLACEMENT (S_ℓ, m)

12 | | fill open locations in set $S_{\ell'}$

13 | | calculate objective function

14 | | $F_{\ell'} = \sum_{n \in S_{\ell'}} f_{well,n}$
    | % WELL SELECTION:

15 | RANK individuals $\{S_1 \ldots S_{Npop}\}$ based on F_ℓ

16 | Keep highest ranked $(Npop/6)$ individuals
    | and renumber them as $\{S_1 \ldots S_{Npop/6}\}$

17 Choose optimal achieved set of well locations

18 $S^* = S_m$ where set S_m satisfies $F_m = \max_j F_j$

6.2.3 Very Fast Simulated Reannealing Algorithm Design

In contrast to GA, where mutations are carried out in parallel, within simulated annealing, a single parameter vector is changed serially, akin to having a single individual in a genetic algorithm. However, every change to the parameter vector must pass the acceptance criteria, or the change is rejected. In the SA algorithm used, first the system parameter (set of selected well locations) and coexistence constraint matrix are created in the same way as they were for the genetic algorithm. During the simulated annealing algorithm, presented in pseudocode as Algorithm 6.3, the size of the parameter change is made according to the current temperature, and both the temperature and parameter size change decrease according to a cooling schedule as the number of iterations advances. As the cooling proceeds, the parameter will move to a local minimum in the objective function. To give the parameter the chance to move out of this minimum, the cooling schedule is periodically reset, a process that is called *reannealing*.

Algorithm 6.3 Pseudocode for Very Fast Simulated Reannealing.

$S^* = \text{VFSR}(d_{min}, N_{well})$

```
 1  S₁ = GREEDY_SET (d_min, N_well) with objective value F*
 2  for i = 1 ... number_of_reanneals
 3      S = S* with objective value F = F*
 4      for j = 1 ... cooling_steps
 5          set temperature parameter Tⱼ
            % select location of replacement well:
 6              for m = 1 ... Nₓ × Nᵧ, m ∉ S
 7              |  δ(m) = Σₙ∈ₛ(Cₘ,ₙ == 0)
 8              find Δ according to cooling schedule Tⱼ
 9              choose well m from the set m ∈ {n | δ(n) = Δ}

10          Sₚ = WELL_REPLACEMENT (S, m)
11          fill open locations in set Sₚ
12          Fₚ = Σₙ∈ₛₚ fₚₑₗₗ,ₙ
            % acceptance criteria:
13              if Fₚ ≥ F
14              |  S = Sₚ & F = Fₚ
15              |  if Fₚ ≥ F*
16              |  |  S* = Sₚ & F* = Fₚ;
17              else
18              |  ΔF = F − Fₚ
19              |  draw random number u ∈ U[0,1]
20              |  if u < exp(− ΔF/ΔT)
21              |  |  S = Sₚ & F = Fₚ
```

In the current work, the system parameter is the selected set of well locations. The size of the parameter change was chosen to be the number of selected well locations that must be deleted in order to accommodate the adding of a new well while meeting the spacing constraints. Therefore, the parameter size change is dependent on which replacement well not in the set is chosen to be added to it. As cooling proceeds, only those wells that do not conflict with many of the existing wells in set S will be considered due to the decreasing temperature. After a replacement well is chosen, it is used in the well replacement operation to a new potential parameter, set S_p. S_p will replace the current set of wells S only if it meets the acceptance criteria. The probability of accepting a set S_p with a lower objective value than the current S also decreases as cooling proceeds, which forces the parameter toward a local minimum toward the end of an annealing cycle. However, replacing S with an S_p having a lower objective value is possible at the beginning of each cooling cycle, helping the parameter to jump out of the local minima in the hope that it can reach a lower minimum.

Within the simulated annealing algorithm, a set of wells with the highest achieved objective value is retained and replaced only when a new set exceeds its objective value. At the start of every reannealing cycle, the parameter to be changed is set to this best set. After either the total number of desired reanneals occurs or the maximum computational time limit is reached, the best set of wells achieved with the highest objective value is returned to the user.

6.3 Optimization via Mathematical Formulations

Mathematical optimization approaches quantitative problems using tools such as linear algebra, calculus, and graph theory. In a sense, it is a more sophisticated method than evolutionary metaheuristics, and, in practice, usually requires bigger and more structured algorithms to solve a given problem. The main advantage of using mathematical optimization and, in our particular case, IP, is that a solution may be proven to be optimal, as opposed to GA, which does not guarantee optimality of the best solution found.

It is common practice in IP to formulate problems by defining an objective function to be maximized (or minimized), subject to constraints that define the problem in question. Here, the formulation of the well placement problem was presented in AlQahtani et al. (2012).

6.4 Optimization Computations

The computations were carried out on a single machine that has a 2.4-GHz quad-core CPU with 32 GB RAM. For tests using IP, the formulations generated were written in MATLAB, and the optimization models were resolved using the GUROBI 5.6 solver with default settings, which were set to the branch-and-cut solution strategy, with absolute gap tolerances set to zero. The other controls used are the same as described in Alqahtani (2013).

Our test grid has $m = n = 100$, and we set the values of D to 6, 8, 10, 12, 14, 16, 18, and 20. For each of these values, we tried to solve the WPP with N set to 10, 20, ..., up to the point at which either the time limit of 3600 seconds was reached or the value of N was not reached, that is, the point at which either method was unable to find a solution that represented N well locations. For the GA, we set $n_g = 1000$, $n_p = 1000$, $n_m = 5$. For the very fast simulated re-annealing (VFSA), we set the cooling steps per anneal $=1000$, the reanneals $=500$, and a search space of 2000 for the GA and VFSA tests.

The values of the P matrix were obtained with the commercial reservoir simulator Eclipse using the quality maps approach (see Da Cruz et al. (1999) for details). The main characteristics of our heterogeneous and anisotropic reservoir are as follows: initial pressure of 4000 psi, porosity of 22%, and horizontal permeability average of 175 mD with standard deviation of 91.1 mD. The distance between the two closest grid points is 300 ft, and the thickness of the reservoir is 75 ft.

The results obtained using GA, VFSA, and IP are shown in two sets. The first set includes performance comparison in terms of objective function values achieved and elapsed computation time. This set of comparisons can be found in Figures 6.2 through 6.9 when $D = 6$, 8, 10, 12, 14, 16, 18, and 20, respectively. Further, the term "Best Sol'n" is used in the first set of figures to represent the value of the best objective function solutions found by any of the methods. (Note: In GA, this value is called the fitness of the best individual, whereas in VFSA and IP, it is called the objective function value of the best solution. We will use the latter expression in our analysis in this chapter and the next.) The expression "Time" shows the computational time in seconds required for the test to finish. We note that, because of the settings used in GUROBI, all solutions obtained using IP are provably optimal precisely when the time is under 3600 seconds.

The second set of results includes variation among the three methods investigated with regard to the number of wells each method placed. This set

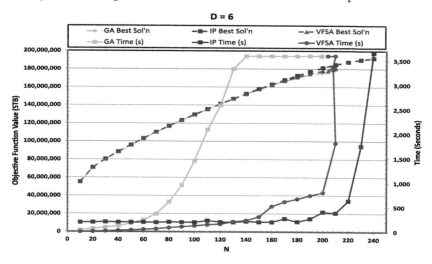

FIGURE 6.2
GA, VFSA, and IP performance comparison in the subject of time and objective function values for $D = 6$.

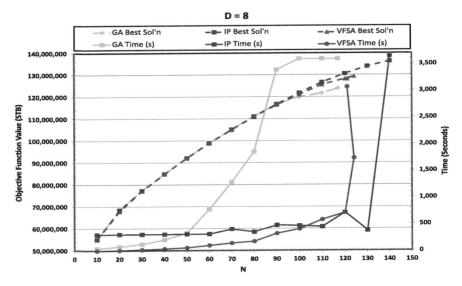

FIGURE 6.3
GA, VFSA, and IP performance comparison in the subject of time and objective function values for $D = 8$.

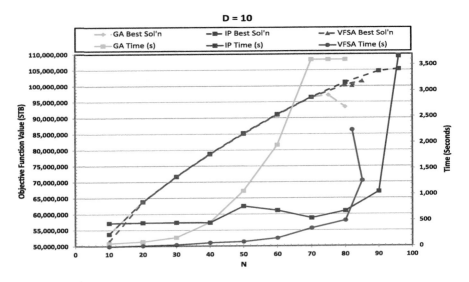

FIGURE 6.4
GA, VFSA, and IP performance comparison in the subject of time and objective function values for $D = 10$.

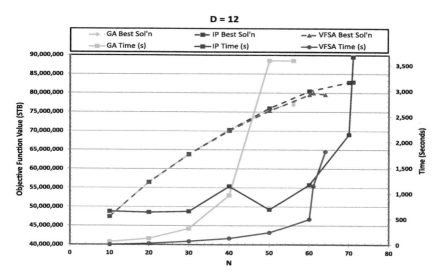

FIGURE 6.5
GA, VFSA, and IP performance comparison in the subject of time and objective function values for $D = 12$.

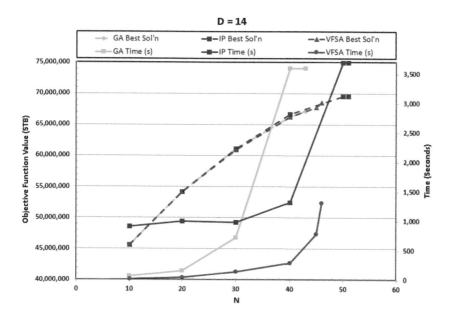

FIGURE 6.6
GA, VFSA, and IP performance comparison in the subject of time and objective function values for $D = 14$.

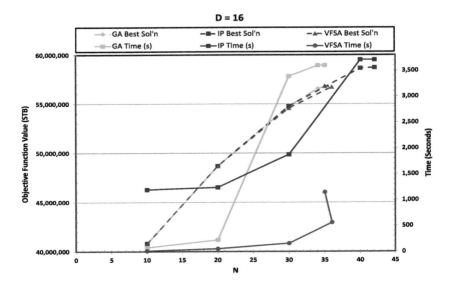

FIGURE 6.7
GA, VFSA, and IP performance comparison in the subject of time and objective function values for $D = 16$.

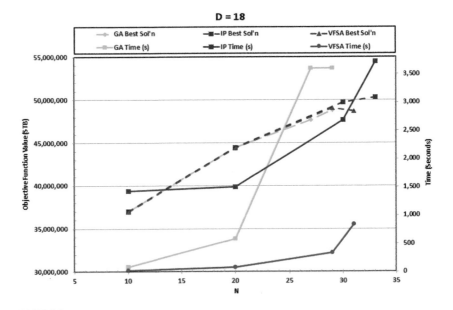

FIGURE 6.8
GA, VFSA, and IP performance comparison in the subject of time and objective function values for $D = 18$.

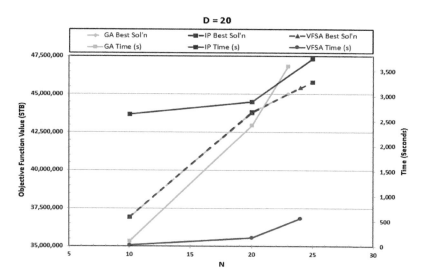

FIGURE 6.9
GA, VFSA, and IP performance comparison in the subject of time and objective function values for $D = 20$.

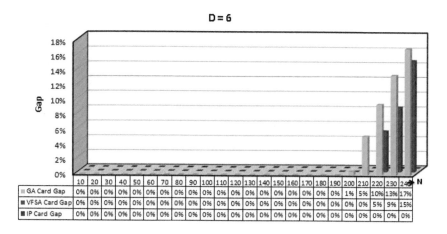

FIGURE 6.10
Bar chart showing the cardinality gaps. The colors blue, red, and green show the cardinality obtained using IP, VFSA, and GA, respectively, for $D = 6$.

shows, in the same figures, the actual number of wells accommodated by each method. The second set of results can be found in the form of flowcharts in Figures 6.10 through 6.17 when $D = 6, 8, 10, 12, 14, 16, 18,$ and 20, respectively.

In the second set, the term "Card Gap" is used to refer to the cardinality constraint, which shows the actual number of wells each method was able to accommodate, relative to the assigned value of N, based on Equation 6.1

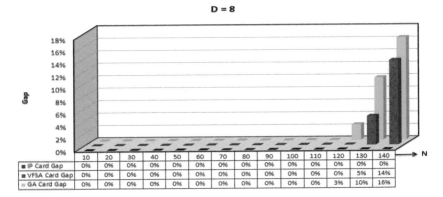

FIGURE 6.11
Bar chart showing the cardinality gaps. The colors blue, red, and green show the cardinality obtained using IP, VFSA, and GA, respectively, for $D = 8$.

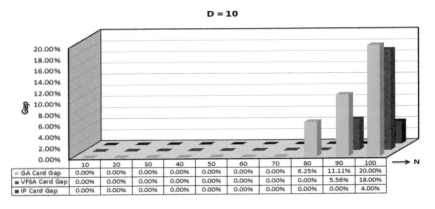

FIGURE 6.12
Bar chart showing the cardinality gaps. The colors blue red, and green show the cardinality obtained using IP, VFSA, and GA, respectively, for $D = 10$.

$$\text{Card Gap} = \frac{N - (\text{GA or VFSA or IP}) \, \text{Card}}{N} \cdot 100 \qquad (6.1)$$

From the first set of figures, it is clear that the difficulty of the problem increases as the values of D and N increase, as illustrated in Figures 6.2 through 6.9. On the IP side, this increase in difficulty is reflected in the time required to solve the instances, which is denoted with solid blue lines. On the GA and VFSA side, both the computational time, denoted with solid green lines for GA and solid red lines for VFSA, and the cardinality gap of the best solution, represented with green bars for GA and red bars for

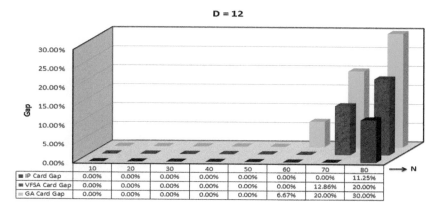

FIGURE 6.13
Bar chart showing the cardinality gaps. The colors blue, red, and green show the cardinality obtained using IP, VFSA, and GA, respectively, for $D = 12$.

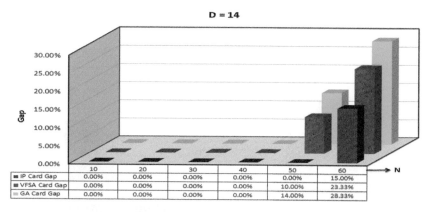

FIGURE 6.14
Bar chart showing the cardinality gaps. The colors blue , red, and green show the cardinality obtained using IP, VFSA, and GA, respectively, for $D = 14$.

VFSA in Figures 6.10 through 6.17, show the increase in the difficulty of the problem. For instance, for $D = 6$ and $N = 210$, the best solution that GA can find has a cardinality gap of 5.24% (i.e., 199), although solutions with a zero cardinality gap (and better objective function values) do exist, as the IP and VFSA results show in Figures 6.2 and 6.10. When $N = 240$, for $D = 6$, the cardinality gaps for GA and VFSA reach 17.1% and 15%, respectively, even though a zero cardinality gap (and better objective function) is noticed, as the IP results indicate in Figures 6.2 and 6.10. Similar situations are observed when $D = 8$, 10, 12, 14, 16, 18, and 20 for $N \geq 120$, 80, 60, 50, 40, 30, and 30, respectively.

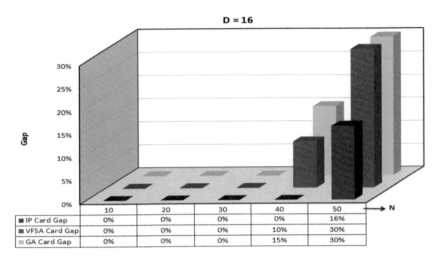

FIGURE 6.15
Bar chart showing the cardinality gaps. The colors blue, red, and green show the cardinality obtained using IP, VFSA, and GA, respectively, for $D = 16$.

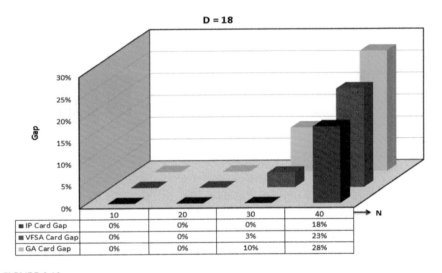

FIGURE 6.16
Bar chart showing the cardinality gaps. The colors blue, red, and green show the cardinality obtained using IP, VFSA, and GA, respectively, for $D = 18$.

Overall, IP was capable of finding better solutions than the GA and VFSA, particularly as the instances increased in difficulty. In every test, the best solution found by IP, which is denoted with dashed blue lines in the first set of figures, had an objective function value at least as good as the one found by the GA, which is denoted with dashed green lines in the same graphs, or

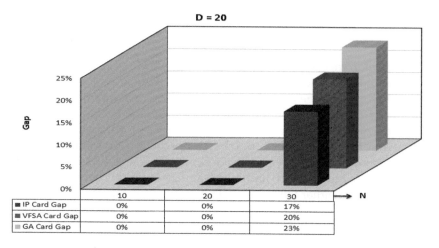

	10	20	30
■ IP Card Gap	0%	0%	17%
■ VFSA Card Gap	0%	0%	20%
▨ GA Card Gap	0%	0%	23%

FIGURE 6.17
Bar chart showing the objective function and cardinality gaps. The colors blue and red show the cardinality obtained using GA and IP, respectively, while the color green displays the objective function difference between the two methods for $D = 20$.

the one found by VFSA, which is denoted with dashed red lines. In 10 cases, the three methods found an optimal solution to the problem. In 4 other instances, the objective function gap between the IP and GA methods was $\leq 1\%$, and in 43 instances, the objective function gap between the IP and VFSA methods was $\leq 1\%$. This result shows that the GA and VFSA can be effective for instances that are less challenging, but for the more difficult instances, IP was notably more successful than either method.

The merit of being faster and finding better solutions is due not only to the differences in approaching the problem (IP vs. VFSA vs. GA), but also a result of the algorithms used to solve the instances. While the developed GA or VFSA algorithms have hundreds of code lines, GUROBI is a state-of-the-art professional solver package. The apparently dashed lines representing the values of the objective functions from the three methods, as in the first set of figures, can go as high as 11% between IP and GA when $D = 10$ and $N = 100$, and 7% between IP and VFSA when $D = 6$ and $N = 240$, which underscores the contrast between the investigated algorithms. For instance, a gap as small as 0.1% can be very difficult to close; thus, finding a suboptimal solution can be a much easier task than finding an optimal one for the GA or VFSA method. Moreover, the fact that GUROBI can guarantee the optimality of a solution is a feature that neither GA nor VFSA has. It is only in comparison studies, such as this one, that the effectiveness of GA or VFSA can be visualized in the finding of optimal solutions. Finally, we note that GUROBI is a multipurpose solver that can handle a vast number of different problems, while our GA and VFSA were developed specifically to tackle the WPP.

6.5 Conclusions

In this chapter, we compared the performance of a GA, VFSA, and IP to solve instances of the problem of placing vertical wells in an $m \times n$ grid that sits on a reservoir, with the objective of extracting the maximum amount of oil from the reservoir. Our results indicate that the GA and VFSA can be effective for easier instances, but lack performance in more difficult instances. In comparison, GUROBI (which takes the IP approach) always found a solution at least as good as that developed by the GA or VFSA, and did so more quickly in many cases. Moreover, IP has the advantage of finding provably optimal solutions, while neither the GA nor VFSA is able to guarantee that a solution is optimal.

The GA presented here is designed and developed differently than the one used in Alqahtani (2013), with minimum alterations for the genetic parameters, which can be sensitized and enhance its performance. The VFSA, likewise, was developed with minimal edits to the annealing parameters in which common values were used, and thus can be improved by sensitizing the annealing parameters or simply changing the algorithm. Similarly, the solution via IP can be made faster by considering different formulations of the problem and tuning some parameters on GUROBI. We opted not to complicate the investigated approaches, keeping them simple without revamping genetic, annealing, and tuning parameters for the GA, VFSA, or GUROBI, respectively.

7

Two-Dimensional Mathematical Optimization Approach for Well Placement and Fracture Design of Shale Reservoirs

7.1 Introduction

The low permeability of shale requires hydraulic fractures to maximize the contact area with the reservoir. The heterogeneity of shale makes finding the optimal zones to fracture challenging. Further complicating the determination of an optimal zone is the heterogeneous presence of quartz, feldspar, carbonate, clays, and organic matter (kerogen) at production zones. Nonetheless, producers must identify "sweet spots" in shales with high levels of hydrocarbons that also easily fracture into complex crack networks with large contact areas. One approach to this challenge maps the geochemical properties of a reservoir using the mineralogical index (Alzahabi et al. 2015c) to guide optimal wells and the fracturability index (Alzahabi et al. 2015b) to guide fracture placement and sequencing.

Wells are commonly placed uniformly with fractures equally spaced and distributed along the well's path. However, this approach does not guarantee optimal production. One explanation for suboptimal placement is that multiple simultaneous fractures can create stress shadowing that can affect the overall reservoir performance by causing some fractures to reorient, resulting in suboptimal production of some fracture stages. To help prevent or minimize stress shadowing, the FI was developed to prioritize and schedule fracture positioning (see Alzahabi et al. 2014).

AlQahtani et al. (2013) demonstrated the superior performance of integer programming, a computational optimization technique for solving constrained mathematical problems, for vertical well placement based on quality index values.

The existing work regarding the placement of wells and fractures in shale and tight rock uses stochastic search optimization techniques. Other techniques are unreliable for optimal placement because of the randomness of reaching optimality and are suboptimal solutions for reaching an objective function. Ma et al. (2015) presented an optimization framework to place horizontal wells and fractures with equal spacing based on testing gradient, gradient-free methods, and the genetic algorithm. Stochastic approaches such as GA and hybrid optimization methods have been used for placing wells (Guyagular and Horne 2001; Yeten et al. 2003). Vasantharajan and Cullick (1997) proposed the combination of quality maps with an IP solution in order to solve the well placement problem. For each case studied, it was found that IP was capable of finding solutions at least as good as those obtained from the genetic algorithm.

Here, we formulate and solve a constrained design problem with an associated objective function to determine the optimum well and fracture placement (for details on optimization methods in oil gas development and usage of IP, see Aronofsky (1983)). The constraints of the design problem model include a minimum distance between fractures and wells. The requirement is enforced using equations expressing these constraints. The objective function works by assigning a value to each feasible solution of the constrained problem based on a linear combination of the decision variables; for example, a solution with a higher objective function value would have greater expected hydrocarbon production. Optimization of the design problem occurs when a maximum (or minimum) objective value is reached and all constraints are respected. A mixed integer program (MIP) is a type of linear program (LP) in which all decision variables must take integral values. The advantage of using IP is that the certificate of optimality is guaranteed in any optimal solution. This approach differs from suboptimal approaches such as GA, which can give optimal solutions but often become trapped in local optima. The presented integrated approach includes geomechanical, geochemical, petrophysical, rock, and fluid properties, and has been shown to be a valuable tool in designing shale reservoir development plans (Alzahabi et al. 2015c). The optimization approach used in this work is different from other fracture design methods such as the technique developed by Economides et al. (2002) in that this approach uses mathematical principles as opposed to an ad-hoc workflow methodology.

Here, we develop a computational approach to design wells and schedule hydraulic fractures to employ the FI index (for details, see the flowchart in Figure 7.1). For years, industry researchers have accepted the zipper fracture (East et al. 2011), modified zipper fracture (Rafiee et al. 2012), and sequential fracturing approach (Uhri et al. 1986) to place fractures in unconventional reservoirs. In this chapter, we present a novel optimization approach for the design of wells and fractures based on the FI map. Minimum well spacing is guaranteed to account for the fracture network dimension constraints. Justification for the fracture sequence is to allow a minimum time span in job implementation and operational purposes, which would guarantee complexity and access to larger surface areas.

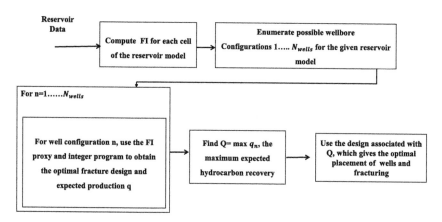

FIGURE 7.1
Flowchart used in optimization.

7.2 Materials

To maximize Equation 7.1 using IP methodology, we use a commercial solver based on branch-and-cut optimization. We have used the GUROBI (Gurobi Team 2016) software package to solve practical instances of the problem and MATLAB in generating instances of the problem.

7.3 Development of the Mathematical Formulation

7.3.1 Objective Function

The goal of well and fracture placement is to maximize the expected total hydrocarbon recovery while respecting the cost and operating constraints. The value of the FI for that cell represents a proxy for the expected hydrocarbon recovery from any cell of the reservoir model. Therefore, the objective function can be best represented by maximizing the total sum of FI values from fractured cells based on the well and fracture placement.

The objective function given in Equation 7.1 states that the goal of the developed scheme is to maximize the sum of the FI values at the points where fractures propagate.

$$\max \sum_{k} \sum_{j \in X^k} \mathrm{FI}_j x_j^k \tag{7.1}$$

7.3.2 Constraints

The following are basic types of constraints (See Figures 7B.1 through 7B.5 for geometric description of the constraints.):

   i. Maximum size of each fracture

   ii. Maximum total cost of all wells and fractures

   iii. Connectivity between each fracture and wellbore

   iv. Spacing between fractures and between fractures and wellbores

   v. Spacing of fractured node j and other fractures from the same wellbore

The values used in the description of this model include:

   i. Reservoir boundary (X_e and Y_e)

   ii. Fracture length and width

The constraints limit the geometry, maximum dimensions, and relative placement of each well and fracture. The size of each fracture could be specified by fracture stage width and length, in terms of the proppant volume and type, or in terms of simulated reservoir volume. In this work, the maximum half-length of each fracture was constrained.

The maximization of the function in Equation 7.1 is subject to the number of well constraints in the following equations:

$$x_j^k + x_i^k \leq 1, \quad \forall i \in \left(M^k \cap E_{12}^j\right), \quad j \in M^k, \quad k \in 1,\ldots,N_w \qquad (7.2)$$

Equation 7.3 states that a minimum spacing of fractures emanating from different wellbores is specified as:

$$x_j^k + x_i^\ell \leq 1, \quad \forall j \in X^k, \quad \forall i \in \left(X^\ell \cap E_{11}^j\right), \quad \forall k, \ell = 1,\ldots,N_w, k \neq \ell \qquad (7.3)$$

Spacing of fractures emanating from different wellbores is constrained as shown in Equation 7.4:

$$x_j^k + x_i^\ell \leq 1, \quad \forall i \in \left(X^\ell \cap E_{11}^j\right), \quad j \in X^k, \quad k \in 1,\ldots,N_w, \ell = 1,\ldots,N_w, \ell \neq k \qquad (7.4)$$

Equation 7.5 constrains the connectivity of the fracture to the wellbore:

$$x_j^k - x_{j^*}^k \leq 0, \quad \forall j \in X^k, \quad \forall k = 1 \ldots N_w \qquad (7.5)$$

Equation 7.6 shows symmetry of the fracture around the wellbore:

$$x_j^k - x_{j+}^k \leq 0, \quad \forall j \in X^k, \quad \forall k = 1 \ldots N_w \tag{7.6}$$

Equation 7.7 constrains the total cost of fracturing:

$$\sum_{k=1}^{N_w} \sum_{j \in X^k} C_j^k x_j^k \leq B \tag{7.7}$$

The maximum fracture length constraint is presented in Equation 7.8:

$$\text{Fracture length}, X_f < \text{distance to drainage boundary}, X_e \tag{7.8}$$

7.3.3 Stress Interference or Shadowing Effect

7.3.3.1 Same Wellbore

Once the first fracture is created, pressure inside the fracture is still high, causing an excess in the resultant stresses that the second fracture in the same well will suffer, sometimes changing propagation orientation. Allowing sufficient spacing and time for implementation of two consecutive fractures would help avoid this phenomenon, which Daneshy (2014) called *intrawell shadowing*. This problem is addressed here by allowing minimum spacing and a certain time span between implemented fractures.

7.3.3.2 Two Adjacent Parallel Wells

The alternation of fractures between two parallel horizontal wells would help reduce the effect of stress shadowing. It would also help in allowing the time span of implementing sequential fractures.

7.3.4 Model Limitation

For the current model, we assume that we have two horizontal parallel wells in the shale model, and the minimum distance between a fracture extending from one well and another wellbore is of a variable input. The maximum length of a half-fracture cannot exceed the well spacing.

7.3.5 Fracture Design Optimization Approach

In order to get the proper formulation, the following parameters are determined by use of the optimization technique:

7.3.5.1 Input Parameters

    i. Fracture dimensions (length, assuming the fracture propagates as a network through the whole fracture stage)

    ii. Fracture spacing

7.3.5.2 Output Parameters

    i. Number of fractures

    ii. Fracture locations

    iii. Optimum scheduling of fracture stages (rank based on FI values)

    iv. Well spacing

Assumption:

- Symmetric planar bidirectional propagation of the fracture from the wellbore occurs perpendicular to the direction of minimum principal stress.

Considerable reduction in the search space for the set of wellbore trajectories is achieved by requiring the wellbores to be drilled in the direction of the minimum horizontal stress. This requirement also allows an iterative process to be proposed. First, enumerate all wellbore configurations that are possible for a given reservoir model, then solve the following linear program for each wellbore configuration, and obtain the optimal design over the entire search space of wellbore configurations. Figure 7.1 depicts this process, which is described next.

7.3.6 Decision Variables

    $x_j^k \in \{0,1\}$, Indicates if cell j is fractured from wellbore k

7.3.6.1 Sets

    $X^k \subset \{1...N\}$, Set of indices of the cells able to be fractured from wellbore k

    $W^k \subset \{1...N\}$, Set of cells containing wellbore k

    $M^k \subset \{1...N\}$, Set of cells directly neighboring wellbore k

    $E_{11}^j = \{\ell \mid \ell = 1...N, d(j,\ell) < d_{11}\}$, Set of cells that violate the minimum spacing d_{11} with cell j

    $E_{12}^j = \{\ell \mid \ell = 1...N, d(j,\ell) < d_{12}\}$, Set of cells that violate minimum spacing d_{12} with cell j

7.3.6.2 Construction of set X^k

    $j \in X^k$ if (1–3) are all true:

1. $j \in 1...N$
2. $d(j,\ell) \leq L_f$ for some $\ell \in W^k$ (cell j is sufficiently close to wellbore k)
3. $d(j,m) \geq d_{fw} \quad \forall m \in W^p, \quad p = 1...N_w, \quad p \neq k$ (cell j is sufficiently distant from all other wellbores)

7.3.7 Indices

$j^+(k)$, Index of cell symmetric to cell j across wellbore k

$j^*(k)$, Index of cell connecting cell j to wellbore k along the plane of minimum horizontal stress

7.3.8 Parameters

$FIj \in [0,1]$, Fracturability index value for cell j of the reservoir

$C_j^k \in [0,1]$, Cost of fracturing cell j from wellbore k

N, Total number of cells in reservoir model

B, Maximum allowable cost of all fracturing

N_w, Number of wellbores (total wells and laterals)

d_{12}, Minimum spacing between fractures emanating from different wellbores

d_{11}, Minimum spacing between fractures emanating the same wellbore

d_{fw}, Minimum spacing between fractures emanating from one wellbore and all other wellbores

L_f, Maximum fracture half-length

In the optimization work presented in this chapter, we assume that the wells are horizontal or very close to horizontal.

7.4 Computational Tests and Results

We have tested this approach by developing multiple test grids based on one real reservoir model dataset obtained from the Permian Basin Wolfcamp shale reservoir. The FI values used to test our optimization approach were derived from data given in Table 7.1, using the methodology developed in Alzahabi et al. (2015b). The flowchart in Figure 7.1 provides the procedure to convert the input (geomechanical properties from reservoir data) to output (a map of optimal placement of wells and fractures). Figure 7.1 describes the process.

TABLE 7.1

Range of Geomechanical and Geochemical Data Ranges for
Building Case Study 1 and Case Study 2 Used for Testing the
Optimization Approach

Parameters	Min. Value	Max. Value
E, psi	0.38 E6	9.75 E6
ν, ratio	0.02	0.38
Calcite, wt. %	0.00	83.0
Quartz, wt. %	6.00	75.0
Pyrite, wt. %	0.00	8.00
Clay, wt. %	3.00	49.0
Density, ρ, g/cc.	2.40	2.71

7.4.1 Case Study 1: 50 × 50 × 1

An implementation of the optimization model was written in MATLAB. In
this implementation, a 2-dimensional reservoir was divided into equally
sized cells. The mathematical optimization-developed model was applied
for each of the two cases.

The first case study is for dimensions $x = y = 50$ cells. (See model parameters
[Table 7B.1] in Appendix B.) We set the values of minimum spacing between
fractures (D) to be 2, 3, 4, and so on cells in spacing. Our main model consists
of a thickness of 1000 feet, where 2500 grid cells were used to represent the
shale reservoir. See Table 7.2.

An example FI map is illustrated in Figure 7.2. A shale gas model as
recommended by the optimizer is shown in a 2D map, which can be generated
in accordance with the methodology described in this chapter. The matrix
map in Figure 7.2 shows that it is not trivial to identify regions with high
potential FI values of the reservoir (regions with high Young's modulus and
low Poisson's ratio). In contrast, the output map of the mathematical IP model
as in Figure 7.3 provides an objective solution to rank the highest potential
cells within the tested mode.

Figure 7.2 also illustrates the tested model of FI populations, where the red
cells are suggested by the developed optimization approach. Those points
are used as the initial points for well and fracturing positioning. Based on

TABLE 7.2

Case Study 1 Description and Properties

Number of Cells	50 × 50 × 1
Well Type	Horizontal
Reservoir Dimension	10,000 × 10,000
Fracture Half-Length, ft.	500,600, 650,700,750
Number of Hydraulic Fractures/Well	0–25
Stage Width	One cell size = 200 ft.

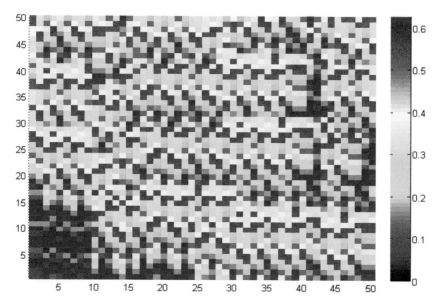

FIGURE 7.2
Fracturability index map based on the tested shale, Case Study 1. The matrix input FI is calculated based on the equation introduced by Alzahabi et al. (2015b).

those points and using the solver, Table 7.3 is generated. Table 7.3 shows the best fracture locations out of 50 possible fracture locations in the shale model. Table 7.3 lists the results of running the developed code and optimum solution by the optimizer for different instances (M1 through M10), Table 7B.2 and 7B.3 demonstrate results of computations for case study 1 of different

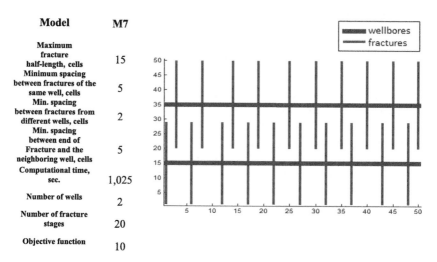

Model	M7
Maximum fracture half-length, cells	15
Minimum spacing between fractures of the same well, cells	5
Min. spacing between fractures from different wells, cells	2
Min. spacing between end of Fracture and the neighboring well, cells	5
Computational time, sec.	1,025
Number of wells	2
Number of fracture stages	20
Objective function	10

FIGURE 7.3
Optimized well and fracture placement.

TABLE 7.3

Summary of Tested Case Study 1 in This Chapter

Input Parameters					Output Results		
Model	Maximum Fracture Half-Length, Cells	Minimum Spacing between Fractures of the Same Well, Cells	Minimum Spacing between Fractures from Different Wells, Cells	Minimum Spacing between End of Fracture and the Neighboring Well, Cells	$t_{optimal}$, sec.	Number of Wells	Number of Fracture Stages
M1	5	2	2	5	246	2	50
M2	10	5	2	5	639	2	20
M3	10	2	2	5	685	2	50
M4	12	5	2	5	1063	2	20
M5	13	5	2	5	1099	2	20
M6	14	5	2	5	1108	2	20
M7	15	5	2	5	1025	2	20
M8	17	5	2	5	1357	2	20
M9	15	2	2	5	1755	2	50
M10	50	9	4	5	17,525	2	11

instances. It compares input parameters for different combinations of input data. It also shows the computational times between different instances of the same model. The computation time increases for a larger fracture half-length and with smaller fracture spacing.

The optimized well and zipper fracture placement in an overlapping frame is shown in Figure 7.3. It also shows the results of testing developed code and the optimum solution obtained by the optimizer for one of the instances. Figure 7.3 shows the geometric overlapping design addressed in our new computational approach to well fracture design for Case Study 1. It is also clear from Figure 7.3 that the two wells and 20 chosen fractures are passing with high fracturability index cells (FI $\geq$ 0.5). This result means that wells and fractures are located in a brittle and good-quality rock.

Figure 7.4 shows the optimal placement for a maximum fracture half-length of 15 and up to 67% overlap. Figure 7.5 shows the optimal placement for a maximum fracture half-length of 12 with no overlap.

The validation of the model was also made possible by use of the second case study described in the next section.

7.4.2 Case Study 2: 80 × 80 × 164

This model consists of an 80 × 80 shale 2D quality map (see model parameters (Table 7B.4) in Appendix B). Figure 7B.6 demnostrates FI

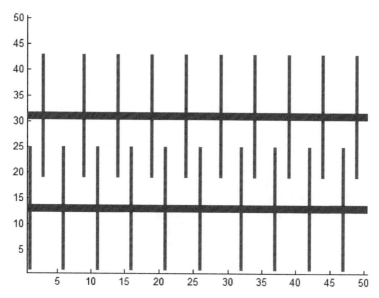

FIGURE 7.4
Optimal placement for a maximum fracture half-length of 15 and up to 67% overlap of fracture stages.

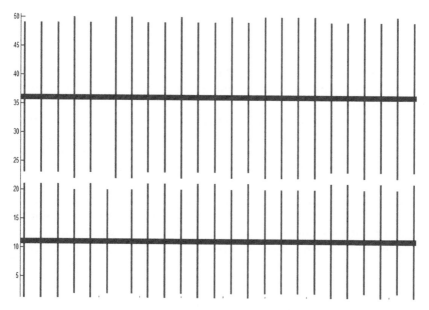

FIGURE 7.5
Optimal placement for a maximum fracture half-length of 10 and no overlap.

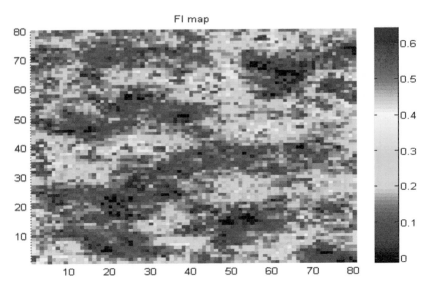

FIGURE 7.6
FI input matrix map of the tested Case Study 2.

matrix map of the chosen layer on a 3D doain of this case study. In this characterization of wells and fracture types over shale derived from the built correlation of FI and plane strain Young's modulus, red regions suggest where the planar fracture will form, blue regions suggest where the rock is more ductile and less likely to support fracturing, and yellow regions suggest where aligned and suboptimal fractures will form. The chosen layer is located at 9400 feet in a zone containing the lowest minimum horizontal stress, higher organic matter content, and higher FI. Figure 7.6 shows the input FI matrix.

Figure 7.7 shows an example of locating wells and fractures for Case Study 2. For a maximum fracture half-length of 17 cells, minimum fracture spacing of 5 cells, and minimum well spacing of 5 cells, the optimum solution is shown in Figure 7.8, where 32 fracture stages in a zipper form are recommended for two wells to drain the reservoir.

One of the aims of this research was to compare an initial location of fractures and a location obtained by the developed optimization algorithm. Figure 7.8 shows a comparison example of locating wells and fractures for Case Study 2, M8 between initial wells and fracture placement and optimized locating of wells and fractures. The optimized case gave a 35% increase in cumulative production versus uniform distribution of wells and fracks.

To validate the optimization technique, we used 2D maps of porosity, TOC, Young's modulus, mineralogical index map (Alzahabi et al. 2015c), permeability map in nanodarcy, and the brittleness index map

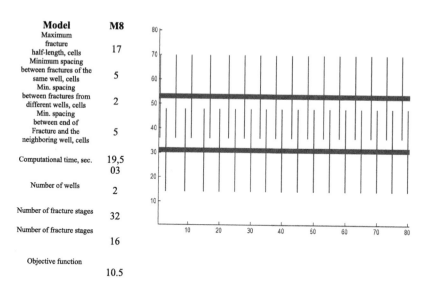

Model	M8
Maximum fracture half-length, cells	17
Minimum spacing between fractures of the same well, cells	5
Min. spacing between fractures from different wells, cells	2
Min. spacing between end of Fracture and the neighboring well, cells	5
Computational time, sec.	19,503
Number of wells	2
Number of fracture stages	32
Number of fracture stages	16
Objective function	10.5

FIGURE 7.7
Well and fracture placement.

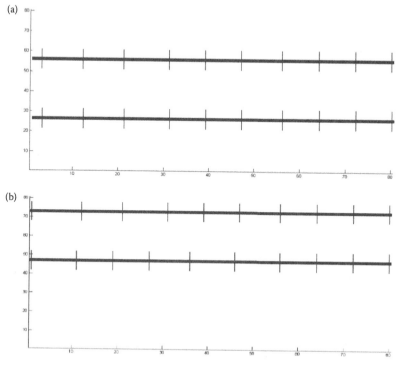

FIGURE 7.8
A comparison of well and fracture placement: (a) top: initial uniform well and fracture placement, (b) bottom: the optimized well and fracture placement.

TABLE 7.4

Summary of Comparison between Different Properties Maps Used for Placing Wells and Fractures

	Porosity Map	TOC Map	Young's Modulus Map	Mineralogical Index	Permeability Map	Brittleness Index, Rickman et al. (2008)
Maximum fracture half-length, cells	17	17	17	17	17	17
Minimum spacing between fractures of the same well, cells	5	5	5	5	5	5
Minimum spacing between fractures from different wells, cells	2	2	2	2	2	2
Minimum spacing between end of fracture and the neighboring well, cells	5	5	5	5	5	5
$t_{optimal}$, sec.	21,290	20,500	21,482	19,652	19,812	19,908
Number of wells	2	2	2	2	2	2
Number of fracture stages	32	32	32	32	32	32
Number of wells	16	16	16	16	16	16
Objective function	6.1	263	289,303,474	33	15,613	19.3

by Rickman et al. (2008) as an input matrix to feed the optimization approach. As shown in Appendix B, we observed a match in placing the wells and fractures between the brittleness index map (Rickman et al. 2008), Young's modulus map, and fracturability index map (Alzahabi et al. 2015b).

Table 7.4 shows testing of our developed approach using different sweet spot identifiers as the input matrix. We note the difference in time of obtaining the optimum solution and objective functional values.

Table 7B.5 shows the best fracture locations out of 80 possible fracture locations in the shale model. Table 7.3 lists the results of running the developed code and optimum solution by the optimizer for different instances (M1 through M9). It compares input parameters for different combinations of input data. It also shows the computational times between different instances of the same model. The computation time increases for a larger fracture half-length and with smaller fracture spacing.

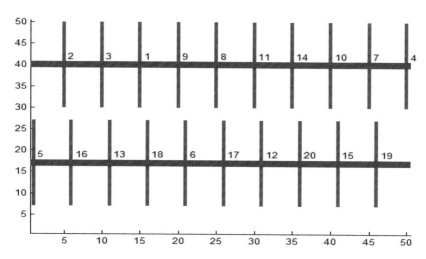

FIGURE 7.9
Rank of fractures according to FI values. A modified zipper fracture is recommended in implementing the design.

7.5 Fracture Stage Sequencing

The sequence of implementing fractures according to the FI matrix values is shown in Figure 7.9. There are two allowable fracture designs, either staggered or overlapping. The numbers rank the possibility of placing fracture stages according to their potential, not according to fracture creation time. An integration with the modified zipper fracture must be implemented. This integration would operationally allow for the implementation of modified zipper fractures guided by FI values as a proxy of sweet spots.

Note that fractures 4 and 5 are not operationally recommended.

7.6 Conclusion and Future Work

In this work, we introduced a new approach for optimizing wells and fracture stage locations, combining the application of FI mapping with the use of mathematical optimization for simultaneous determination of optimal well design and fracture placement. The new technique, using an IP approach with a linear objective function and equality and inequality constraints, can also produce overlapping or zipper fracture-type designs for enhancing complexity. Example implementations of this method are demonstrated for the design of the placement of multiple wells and fractures in a reservoir based on field data for a shale reservoir in Permian Basin. The example case studies show that the optimization technique can produce designs with fractures that are either equally spaced or not, depending on the particular set of reservoir data and constraints considered.

The newly proposed integer programming approach can also be used with other sweet-spot identification techniques. The introduced method relies on the use of the FI for shale quality mapping, providing a link between the objective function to be optimized and the relative ease of creating fracture stages based upon the local reservoir geomechanical properties. Coupling efficient proxies and the interactive process of well and fracture placement reduces the need for many reservoir simulation runs or engineering experience in placing wells and fractures.

For future research, we suggest the extension of this technique to the placement of wells and fractures in unconventional reservoirs within a 3D search domain. Such a study would allow for the optimal design to be determined even in cases where the quality variation cannot be reduced to a 2D model.

Acknowledgments

The authors thank Schlumberger for providing the Petrel software used for modeling the shale reservoir in this chapter.

Appendices

Appendix A: Abbreviations

FI	Fracturability index
IP	Integer programming
E	Young's modulus, psi
E'	Plane strain modulus, psi
E'_n	Normalized plane strain modulus
ν	Poisson's ratio
ρ	Density, lb./ft^3
TOC	Total organic carbon content
MBI	Mineralogical brittleness index
$x_{(i,j)}$	X_Y location of a fracture in the reservoir represents the location (i, j) in the shale formation grid

X	Coordinate axis along well path, ft.
Y	Coordinate axis along fracture path, ft.
BHP	Bottomhole pressure, psi
q_g	Gas flow rate, Mscf/d
L_f	Fracture half-length, ft.
P_i	Initial reservoir pressure, psi
L	Well lateral length, ft.
$\Delta X, \Delta Y$	Model grid dimensions in x and y directions, ft.
P_{net}	Net pressure, psi
X_e, Y_e	Rectangular reservoir shape dimensions in x and y directions, ft.
W	Horizontal well
F	Fracture stage
D_{min}	Minimum well spacing, ft.
R_e	Drainage radius
Time	Time of a single run to obtain optimum solution
Max_{Lfrac}	Maximum fracture half-length
$D_{minfrac}11$	Minimum distance between two fractures extending from different wellbores
$D_{minfrac}12$	Minimum distance between two fractures extending from different wellbores
$D_{min}f_w$	Minimum distance between a fracture extending from one well and another wellbore
$t_{optimal}$	A time in seconds for obtaining optimum solution
N_{wells}	Optimum suggested number of wells
$N_{fractures}$	Optimum suggested number of fracture stages
$N_{fractures}$/well	Optimum suggested number of fracture stages per well

Appendix B: Data Ranges of Main Properties for Both Reservoirs

TABLE 7B.1

Case Study 1 Model Parameters

Cell size $\Delta X, \Delta Y$, ft.	200,200
Lateral length, ft.	10,000
Initial reservoir pressure, psi	5400
Pay zone, ft.	900
L_f, ft.	250–2500
Depth to the top of the formation	6105
c_f, psi^{-1}	3E-06
Porosity, %	10
No. of fractures/well	0–25
Min. fracture spacing	200 ft.

TABLE 7B.2

Results of Computation for Case Study 1

Model	M1
Maximum fracture half-length, cells	5
Minimum spacing between fractures of the same well, cells	2
Minimum spacing between fractures from different wells, cells	2
Minimum spacing between end of fracture and the neighboring well, cells	5
Computational time, sec.	246
Number of wells	2
Number of fracture stages	50
Number of fracture stages/well	25

Model	M3
Maximum fracture half-length, cells	10
Minimum spacing between fractures of the same well, cells	2
Minimum spacing between fractures from different wells, cells	2
Minimum spacing between end of fracture and the neighboring well, cells	5
Computational time, sec.	685
Number of wells	2
Number of fracture stages	50
Number of fracture stages/well	25

Model	M4
Maximum fracture half-length, cells	12
Minimum spacing between fractures of the same well, cells	5
Minimum spacing between fractures from different wells, cells	2
Minimum spacing between end of fracture and the neighboring well, cells	5
Computational time, sec.	1063
Number of wells	2
Number of fracture stages	20
Number of fracture stages/well	10

(Continued)

TABLE 7B.2 (*Continued*)

Results of Computation for Case Study 1

Model	M5
Maximum fracture half-length, cells	13
Minimum spacing between fractures of the same well, cells	5
Minimum spacing between fractures from different wells, cells	2
Minimum spacing between end of fracture and the neighboring well, cells	5
Computational time, sec.	1099
Number of wells	2
Number of fracture stages	20
Number of fracture stages/well	10

Model	M6
Maximum fracture half-length, cells	14
Minimum spacing between fractures of the same well, cells	5
Minimum spacing between fractures from different wells, cells	2
Minimum spacing between end of fracture and the neighboring well, cells	5
Computational time, sec.	1108
Number of wells	2
Number of fracture stages	20
Number of fracture stages/well	10

Model	M8
Maximum fracture half-length, cells	17
Minimum spacing between fractures of the same well, cells	5
Minimum spacing between fractures from different wells, cells	2
Minimum spacing between end of fracture and the neighboring well, cells	5
Computational time, sec.	1357
Number of wells	2
Number of fracture stages	20
Number of fracture stages/well	10

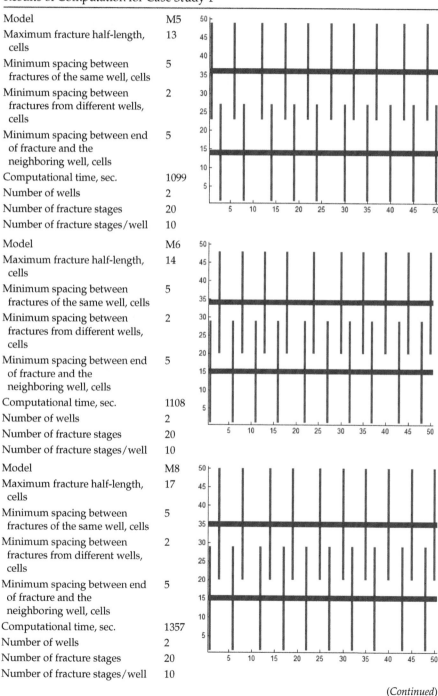

(*Continued*)

TABLE 7B.2 (*Continued*)

Results of Computation for Case Study 1

Model	M9
Maximum fracture half-length, cells	15
Minimum spacing between fractures of the same well, cells	2
Minimum spacing between fractures from different wells, cells, cells	2
Minimum spacing between end of fracture and the neighboring well	5
Computational time, sec.	1755
Number of wells	2
Number of fracture stages	50
Number of fracture stages/well	25
Model	M10
Maximum fracture half-length, cells	50
Minimum spacing between fractures of the same well, cells	9
Minimum spacing between fractures from different wells, cells	4
Minimum spacing between end of fracture and the neighboring well, cells	5
Computational time, sec.	17,525
Number of wells	2
Number of fracture stages	11
Number of fracture stages/well	5–6

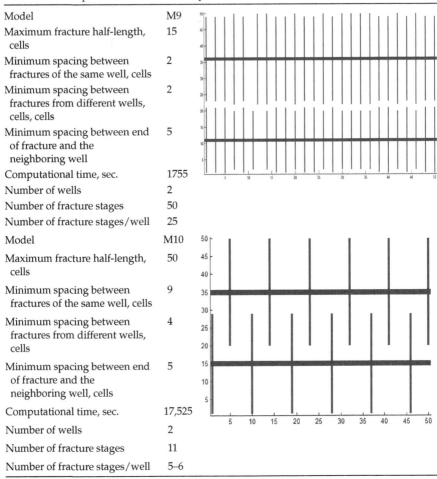

TABLE 7B.3

Results of Computation for Case Study 2

Model	M1	
Maximum fracture half-length, cells	5	
Minimum spacing between fractures of the same well, cells	2	
Minimum spacing between fractures from different wells, cells	2	
Minimum spacing between end of fracture and the neighboring well, cells	5	
Computational time, sec.	1368	
Number of wells	2	
Number of fracture stages	80	
Number of fracture stages/well	40	
Objective function	24.8	
Model	M2	
Maximum fracture half-length, cells	10	
Minimum spacing between fractures of the same well, cells	5	
Minimum spacing between fractures from different wells, cells	2	
Minimum spacing between end of fracture and the neighboring well, cells	5	
Computational time, sec.	4877	
Number of wells	2	
Number of fracture stages	32	
Number of fracture stages/well	16	
Objective function	10.5	

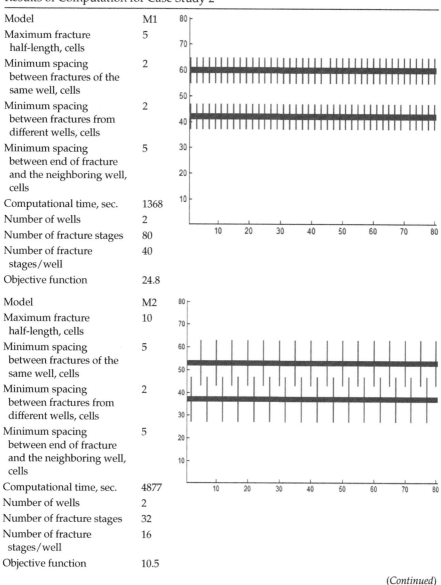

(Continued)

TABLE 7B.3 (*Continued*)

Results of Computation for Case Study 2

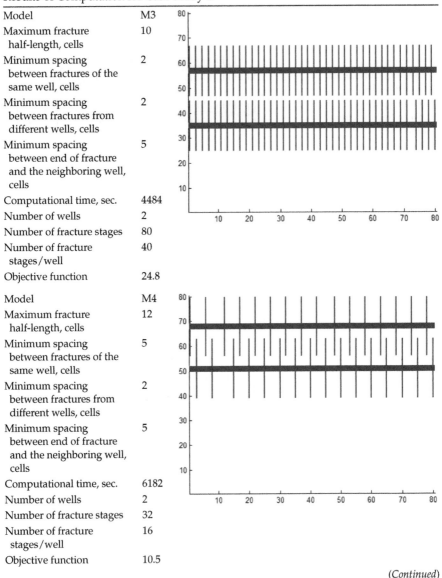

Model	M3
Maximum fracture half-length, cells	10
Minimum spacing between fractures of the same well, cells	2
Minimum spacing between fractures from different wells, cells	2
Minimum spacing between end of fracture and the neighboring well, cells	5
Computational time, sec.	4484
Number of wells	2
Number of fracture stages	80
Number of fracture stages/well	40
Objective function	24.8
Model	M4
Maximum fracture half-length, cells	12
Minimum spacing between fractures of the same well, cells	5
Minimum spacing between fractures from different wells, cells	2
Minimum spacing between end of fracture and the neighboring well, cells	5
Computational time, sec.	6182
Number of wells	2
Number of fracture stages	32
Number of fracture stages/well	16
Objective function	10.5

(*Continued*)

TABLE 7B.3 (*Continued*)

Results of Computation for Case Study 2

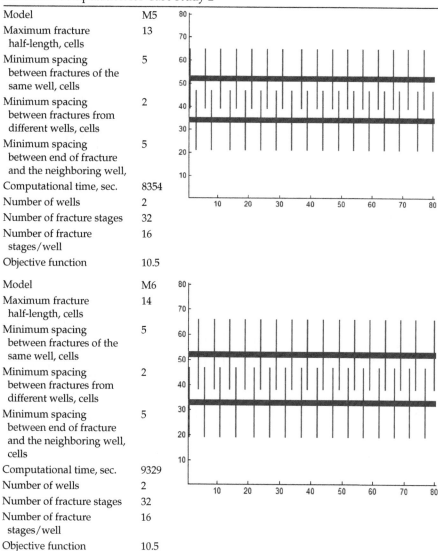

Model	M5
Maximum fracture half-length, cells	13
Minimum spacing between fractures of the same well, cells	5
Minimum spacing between fractures from different wells, cells	2
Minimum spacing between end of fracture and the neighboring well,	5
Computational time, sec.	8354
Number of wells	2
Number of fracture stages	32
Number of fracture stages/well	16
Objective function	10.5

Model	M6
Maximum fracture half-length, cells	14
Minimum spacing between fractures of the same well, cells	5
Minimum spacing between fractures from different wells, cells	2
Minimum spacing between end of fracture and the neighboring well, cells	5
Computational time, sec.	9329
Number of wells	2
Number of fracture stages	32
Number of fracture stages/well	16
Objective function	10.5

(*Continued*)

TABLE 7B.3 (*Continued*)

Results of Computation for Case Study 2

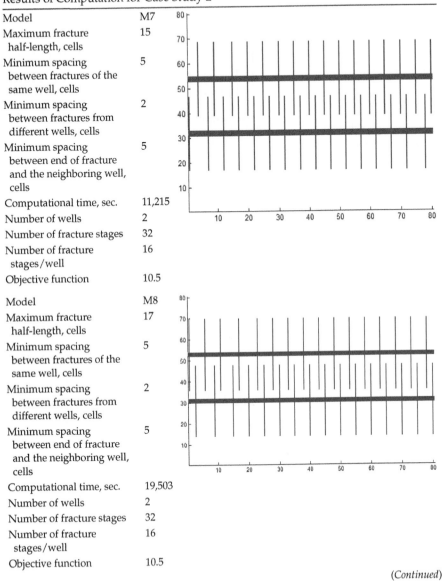

Model	M7
Maximum fracture half-length, cells	15
Minimum spacing between fractures of the same well, cells	5
Minimum spacing between fractures from different wells, cells	2
Minimum spacing between end of fracture and the neighboring well, cells	5
Computational time, sec.	11,215
Number of wells	2
Number of fracture stages	32
Number of fracture stages/well	16
Objective function	10.5
Model	M8
Maximum fracture half-length, cells	17
Minimum spacing between fractures of the same well, cells	5
Minimum spacing between fractures from different wells, cells	2
Minimum spacing between end of fracture and the neighboring well, cells	5
Computational time, sec.	19,503
Number of wells	2
Number of fracture stages	32
Number of fracture stages/well	16
Objective function	10.5

(*Continued*)

TABLE 7B.3 (*Continued*)

Results of Computation for Case Study 2

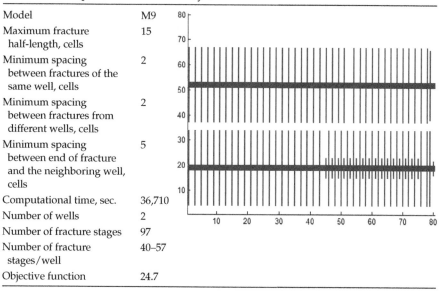

Model	M9
Maximum fracture half-length, cells	15
Minimum spacing between fractures of the same well, cells	2
Minimum spacing between fractures from different wells, cells	2
Minimum spacing between end of fracture and the neighboring well, cells	5
Computational time, sec.	36,710
Number of wells	2
Number of fracture stages	97
Number of fracture stages/well	40–57
Objective function	24.7

TABLE 7B.4

Case Study 2 Model Parameters

Cell size ΔX, ΔY, ft.	25,25
Lateral length, ft.	10,000
Initial reservoir pressure, psi	5400
Pay zone, ft.	900
X_f, ft.	125–1250
Depth to the top of the formation, ft.	7900
c_f, psi^{-1}	3E-06
Porosity, %	10
No. of fractures	16–40
Fracture spacing, ft.	50,125,225

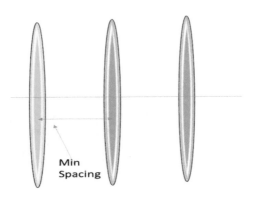

FIGURE 7B.1
Fracture stages and spacing between stages.

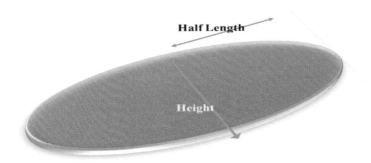

FIGURE 7B.2
One fracture stage main parameters.

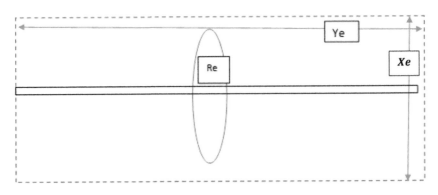

FIGURE 7B.3
Reservoir model boundary constraints.

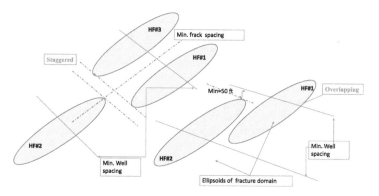

FIGURE 7B.4
Two allowable fracture designs in our mathematical optimization model. Upper left: staggered fracture scheme; upper right is overlapping fracture scheme.

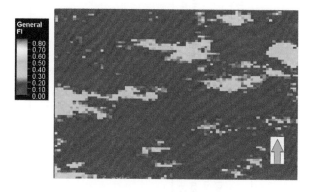

FIGURE 7B.5
FI map of actual chosen reservoir layer for Case Study 2.

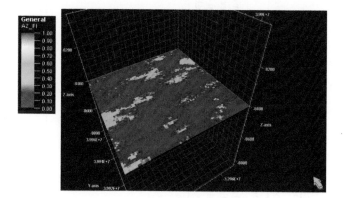

FIGURE 7B.6
FI matrix map of the chosen layer in a 3D domain of Case Study 2.

Table 7B.5 shows a summary of tested Case Study 2 in this chapter.

TABLE 7B.5

Summary of Tested Case Study 2 in This Chapter

	Input Parameters				Output Results			
Model	Maximum Fracture Half-Length, Cells	Minimum Spacing between Fractures of the Same Well, Cells	Minimum Spacing between Fractures from Different Wells, Cells	Minimum Spacing between End of Fracture and the Neighboring Well, Cells	Computational Time, Sec.	Number of Wells	Number of Fracture Stages	Objective Function
M1	5	2	2	5	1368	2	80	24.8
M2	10	5	2	5	4877	2	32	10.5
M3	10	2	2	5	4484	2	80	24.8
M4	12	5	2	5	6182	2	32	10.5
M5	13	5	2	5	8354	2	32	10.5
M6	14	5	2	5	9329	2	32	10.5
M7	15	5	2	5	11,215	2	32	10.5
M8	17	5	2	5	19,503	2	32	10.5
M9	15	2	2	5	36,710	2	97	24.7

Figure 7B.7 shows a comparison case study of Case Study 2 M8 for all maps as a tool for locating wells and fractures. There is a perfect match in placing wells and fractures between the brittleness index map (Rickman et al. 2008) and Young's modulus map. There is also a second match in placing wells and fractures using permeability, mineralogical index map, TOC map, and porosity map.

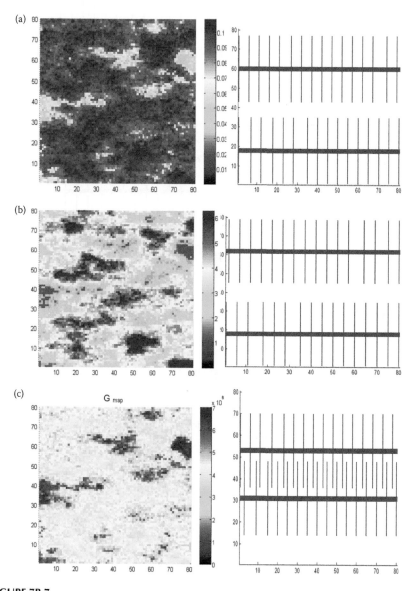

FIGURE 7B.7
(a) through (f) Comparison between different indicators for locating wells and fractures using the developed mathematical approach. *(Continued)*

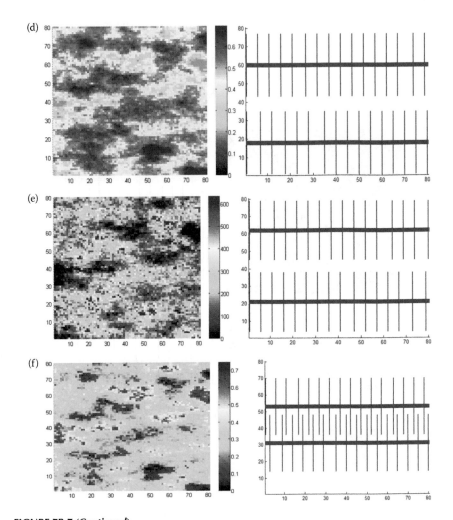

FIGURE 7B.7 (*Continued*)
(a) through (f) Comparison between different indicators for locating wells and fractures using the developed mathematical approach.

8

Multigrid Fracture-Stimulated Reservoir Volume Mapping Coupled with a Novel Mathematical Optimization Approach to Shale Reservoir Well and Fracture Design

8.1 Introduction

SRV has a long history of use in defining the effect of fracturing in shales. Substantial evidence from sonic logs and production data from shale wells support that certain segments of wells make up 70% of the total production of wells. This chapter presents a concept for identifying SRV in shale rock. Creating hydraulic fractures leads to fracture network growth. Fracture growth interaction with existing natural fractures causes complexity. Complexity is a resultant network of induced and existing fractures. SRV is used to account for this resultant complexity. These complex networks have a substantial impact on well performance in shale and tight rocks. The shape of SRV can be predicted from stimulated and shear propped fractures, while the volume can be correlated with fracture network length.

The size of the SRV is correlated to treatment volume. Figure 8.1 shows the relationship between treatment volume and fracture network length for five vertical Barnett shale wells, modified after Fisher et al. (2002).

Mayerhofer et al. (2008) introduced the SRV concept as a 3D size of a created fracture network and defined SRV as a correlation parameter for well performance. Mayerhofer et al. (2010a) linked SRV with well performance of shale reservoirs. A direct relationship is demonstrable between fracture network length and SRV, as shown in Figure 8.1 for Barnett shale wells (modified after Fisher et al. (2002)).

Anderson et al. (2010) defined SRV linked to the horizontal well by stimulated reservoir width, areal extent, and fracture half-length. Zhou and Sill (2013) introduced a method for identifying anisotropic regions in unconventional hydrocarbon reservoirs. Anisotropy can be indicative of sweet-spot zones for fracturing and drilling a productive well. Seismic amplitude data from receivers along two orthogonal lines radiating from a seismic source is used.

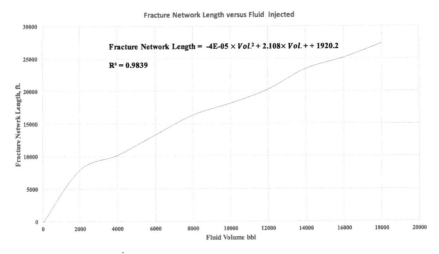

FIGURE 8.1
Fracture network length as a function of fluid volume injected for four vertical wells of Barnett. (Modified after Fisher, M.K. et al. 2002. Integrating Fracture Mapping Technologies to Optimize Stimulations in the Barnett Shale. *Presented at the SPE Annual Technical Conference and Exhibition, San Antonio, TX, September 29–October 2.* doi:10.2118/77411-MS.)

Sil et al. (2013) introduced a method to calculate fracture parameters from common well log data. Fracture parameters can indicate sweet-spot zones in unconventional rock. Microseismic mapping is currently used to map SRV in shale rocks. It is also used as a tool to diagnose the effectiveness of the created hydraulic fractures, especially in multistage fracturing in horizontal wells. Zhang et al. (2015) introduced the SRV equation as follows:

$$SRV = SRA \times H_f = \sum_{i=1}^{n} A_i H_p, \tag{8.1}$$

where H_f is the fracture height; SRA is the stimulated reservoir area; and H_p is the reservoir thickness. One limitation is implicit: it seems to be valid only when $h_p = h_f$.

Cheng et al. (2015) presented an SRV formula as a function of average fracture length, average height of fractures, maximum number of fractures, maximum horizontal stress, and minimum horizontal stress as follows:

$$SRV = \sum_{1}^{n_c} 4\pi x_i^2 h_i \frac{\sigma_H / \sigma_h}{\left(1 + \sigma_H / \sigma_h\right)^2} \tag{8.2}$$

Here
x_i = Average length of the fractures in the ith cluster
h_i = Average height of the fractures in the ith cluster, m

n_c = Number of clusters
σ_H = Maximum horizontal in-situ stress, N/m²
σ_h = Minimum horizontal in-situ stress, N/m²

Additionally, a linear programming–based approach to mathematically optimize SRV is presented. This newly developed optimization approach improves the placement of fractures in quantifiably better zones in shale reservoirs to guarantee optimality of the reservoir development plan given the available data and modeling constraints. This approach will be useful in pad drilling and development of applications applied to shale formations.

This work will lead to the global optimization of the placement of surface pads, location and design of wells attached to the pads, and location of the fractures (SRV) throughout the wells. This design will also take into account other practical design constraints, including length of wells, number of wells associated with a pad, numerous overlap constraints inherent in unconventional gas and oil well development, and so on. The development will be optimized based on maximization of the FI values explored by the final network, and will be constrained by the previously mentioned considerations, as well as a global maximum number of wells and a maximum development budget. In addition, the mathematical framework allows for easy extensibility to other constraints and can be customized based on the problem constraints.

Fracture parameters such as half-length, azimuth, width, and height can be used extensively in the fracture modeling process. Fracture geometry can be modeled instead through SRV, as an estimated fracture volume can give a better description of fracture parameters. SRV is represented by a grid-like geometry that has a value of FI greater than the predefined threshold of 0.5. The SRV consists of a group of cells, as shown in Figure 8.2, and SRV varies from one stage to the next. The next section details the geometric interpretation of the SRV representation, that is, a group of grids characterized by high values of FI.

Before entering into discussion of the mathematical definition, it is necessary to set the objective of developing a method to predict SRV location and number using an input map of FIs. Such a method, when coupled with mathematically developed code, could help in exploitation of the shale resource with the minimum number of wells and number of fracture stages.

The chapter aims at:

- Planning and automating an optimum well path and optimum fracture design in shale and tight formation
- Establishing a process of choice of maximum SRV for future initiation of fractures
- Finding the optimum number of fracture stages
- Optimizing the number of SRVs

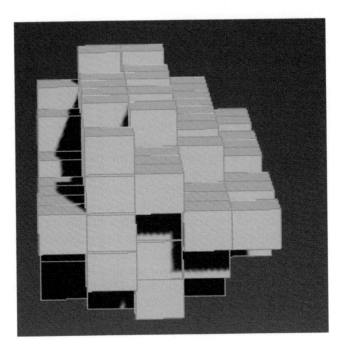

FIGURE 8.2
Multigrid-based SRV.

A previous patent identified an index to help prioritize fracture position and scheduling. Mathematical optimization using integer programming proved its superior performance in vertical well placement (for details on its performance, see Alqahtani et al. (2013)).

Computational concepts such as dynamic programming and graph theory can be useful in exploration of algorithms applied to a wide range of oil and gas optimization topics, the most important part of which is the computational methods used to solve them so that an optimum placement can be obtained. Since the problem is a mathematically based method, the problem definition is first outlined in the next section.

8.2 Problem Definition and Modeling

8.2.1 Geometric Interpretation

8.2.1.1 Fracture Geometry

Hydraulic fracture geometry dimensions may be calculated using analytical approaches based on net pressure and fluid and rock properties. Another common approach is microseismic monitoring, which fits a rectangular box to the microseismic event locations along the horizontal well path. SRV can be estimated based on the volume of the rectangular or principal component box, or by summing a series of volumetric boxes (e.g., Mayerhofer et al. (2008)). For a realistic approach, SRV is used as a representation of fractures connected to the wells. A stimulated approximation of grids is used to represent the hydraulic fracture.

8.2.1.2 The Developed Model Flowchart

In this chapter, an approach for placing surface well pads and fractures in shale rock is shown in Figure 8.3. The solution is presented in Figures 8.3 and 8.4 as follows.

Figure 8.5a and b illustrate the two allowable designs for placing fractures and then SRVs.

8.2.1.3 Well and Fracture Design Vector Components

1. Number of transverse fracture stages per well (5–50).
2. Number of wells of single pad (5–40).
3. Number of perforations per stage (1–6).
4. Length of horizontal well (6000–10,000 ft.).
5. Half-length of fracture (200–600 ft.).
6. Spacing (wells, fractures) (500–1600 ft.).
7. Pay zone thickness (200–1000 ft.).
8. Reservoir boundary dimensions (Ye, Xe) (rectangular shape).
9. Variable stimulated reservoir volume (VSRV) with different variable conductivity.
10. The formation is heterogeneous.
11. The transverse fractures fully penetrate the majority of formations except 10 ft. from the boundary. Fractures are contained within the formation.
12. Multiple transverse fractures are not identical (differencing in dimensionless fracture conductivity and fracture-propped characteristics such as length and network width).

8.3 Development of a New Mathematical Model

In this section, a description of the mathematical model used to solve the above-mentioned problem is given, consisting of an introduction of

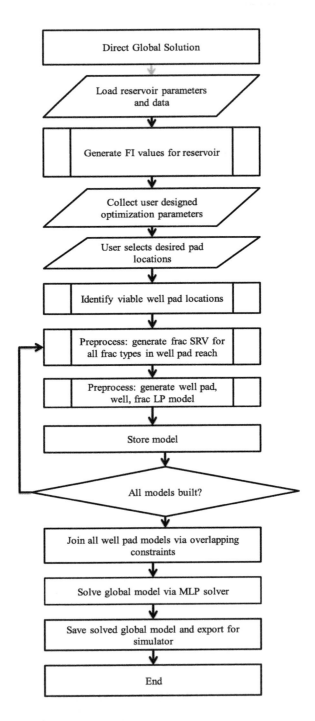

FIGURE 8.3
The first recommended flowchart to be used as a utility in the optimization process shows the interface between pad design and SRVs.

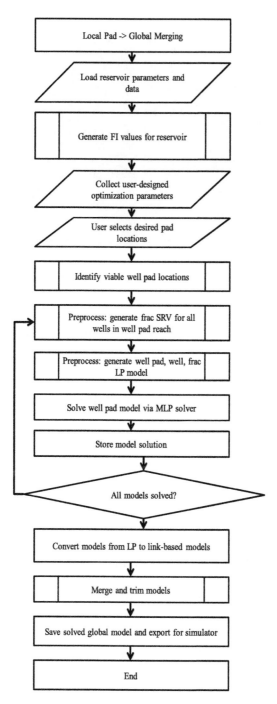

FIGURE 8.4
Second recommended approach of optimization process.

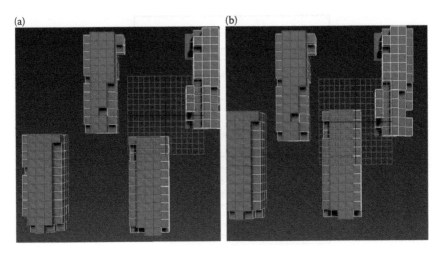

FIGURE 8.5
Four different SRVs are located, whether (a) staggered or staggered and (b) overlapping designs.

methodology, objective function, the essential sets, variables, and constant parameters followed by presentation of the optimization procedure.

8.3.1 Methodology

The following describes the numerical formulation of the newly developed technology.

8.3.2 Objective Function

The problem can be formulated as follows

$$\max \sum A_{mn} X_{mn} + \sum F_k Y_k \tag{8.3}$$

An alternative objective function is to maximize the net income obtained from unconventional reservoirs. The net income is calculated as the difference between total income from total hydrocarbons produced and the total capital and operating expenses, including optimum wells and fractures.

8.3.3 Assumptions and Constraints Considered in the Mathematical Model

Sets, variables, and decision variables are assumed as follows; for a detailed explanation, see geometric interpretation of parameters in Appendix C.
    Where

8.3.3.1 Sets

1. A_{mn}, total FI values unlocked by the stage from node m to node n
2. F_k, total FI values unlocked by fracturing at node k
3. L_{mn}, length of stage from node m to node n
4. $E_I(n)$, edges inbound to node n
5. $E_O(n)$, edges outbound from node n
6. $S(n)$, set of stages that connect to or pass through node n, represented as edges in the network
7. $\Psi(n)$, set of starting nodes mutually exclusive with w
8. E, set of valid edges for the network
9. $\Lambda(m, n)$, set of fractures accessible from the stage between node m and node n
10. $\Phi(n)$, set of nodes whose fracture SRV would intersect the fracture SRV of node n
11. $\Omega(n)$, set of stages intersected or interfered with by fracturing node n

8.3.3.2 Variables

$P_{i,j,k}$ total FI values unlocked by fracturing or placing well at nodes i, j, and k

8.3.3.3 Decision Variables

$w_{i,j,k}^m$ binary variable equal to 1 if well m goes through node i, j, k ($m = 1\ldots$ maxwell)

$f_{i,j,k}^{m,n}$ binary variable equal to 1 if fracture n, extending from well m, goes through node i, j, k ($n = 1\ldots$ maxfrac)

W_n, well origin for single well originating at node n, continuous

X_{mn}, flow of connections from node m to node n, continuous

Y_n, fracturing of node n, continuous

S_{nm}, usage of stage connecting node m to node n, binary

8.3.3.4 Extended Sets

$\Psi_w(m, i, j, k)$, set of nodes that cannot have well if $w_{i,j,k}^m = 1$ (8.4)

$\Psi_f(m, n, i, j, k)$, set of nodes that cannot have well if $f_{i,j,k}^{m,n} = 1$ (8.5)

$\Omega_w(m, i, j, k)$, set of nodes that cannot have well if $w_{i,j,k}^m = 1$ (8.6)
due to constraints on well orientation angle

$\Omega_f(i,j,k)$, set of nodes neighboring node (i,j,k) along the plane of minimum stress

8.3.3.5 Constant Parameters

1. *MinSpace:*	Minimum spacing between fractures
2. *MinStageSpace:*	Minimum spacing between stages
3. *MaxStageSpace:*	Maximum spacing between stages
4. *MaxLength:*	Maximum allowed length for a single well
5. *MaxStages:*	Maximum number of active stages in a well
6. *MaxFPS:*	Maximum number of fractures per stage
7. *MaxWells:*	Maximum number of wells per pad
8. *MaxPads:*	Maximum number of pads per reservoir
9. *MaxPadCost:*	Maximum cost for global development
10. *MaxCost:*	Maximum cost for global development
11. *Thickness:*	Reservoir thickness
12. *WC:*	Cost for initial construction of a well originating from a pad
13. PC_{mn}:	Cost for construction of a well segment from node m to n

8.3.3.6 Constraints

The problem is subject to the following constraints:

1. Fracture half-length $(X_f) < 0.9\,(D1 + D2)$ (8.7)
2. The fracture propagates only along the predefined
   direction of minimum stress at each location   (8.8)
3. Fracture half-length $(X_f) < X_e$ and $X_f < 500$ ft.   (8.9)
4. There is no fracturing in angled paths.
5. $S_{nm} \in \{0,1\}$ [active stage variable, may not be used]   (8.10)
6. $0 \le W_{\alpha,n} \le 1$ (8.11)
7. $0 \le X_{mn} \le 1$ (8.12)
8. $0 \le Y_k \le 1$ (8.13)

9. $$\sum_{p(n,p) \in EO(n)} X_{np} - \sum_{m(m,n) \in EI(n)} X_{mn} \le 0 \text{ [flow conservation with no sink]} \quad (8.14)$$

10. $$\sum X_{m',n'} < 1,\, (m', n') \in S(n) \text{ [nonoverlap constraint]} \quad (8.15)$$

11. $$\sum X_{w,n'} \le 1,\, n' \in \Psi(n) \text{ [nonoverlap constraint]} \quad (8.16)$$

12. $$\sum_{(i,j) \in E} X_{ij} * L_{ij} \le Max\,Length \text{ [constrains length of well]} \quad (8.17)$$

13. $$\sum_{k} Y_k \le MaxStages \text{ [constrains number of stages in well]} \quad (8.18)$$

14. $\sum_{k} W_{a,k} \leq \text{MaxWells}$ [constrains number of wells per pad]   (8.19)

15. $X_{mn} * \text{MaxFPS} + \sum_{k \in \Lambda(m,n)} Y_k \leq 0$ [access to fractures for active

    stages; limited to straight, level stages]   (8.20)

16. $\sum_{k \in \Phi(n)} Y_k \leq 1$ [nonoverlap of fracture/SRV]   (8.21)

17. $\sum_{k \, \text{dist}(n,k) < \text{MinSpace}} Y_k \leq 1$ [minimum spacing between two fractures,

    focused on node n]   (8.22)

18. $|\Omega(n)| * Y_n + \sum_{(i,j) \in \Omega(n)} X_{ij} \leq |\Omega(n)|$ [nonoverlap of fracture

    SRV for node k with stages that the SRV would intersect or be
    too close to (within 50 ft.)]   (8.23)

19. $WC \sum_{k} W_{a,k} + \sum_{(i,j) \in E} X_{ij} * PC_{ij} + FC \sum_{k} Y_k \leq \text{MaxCost}$ (8.24)

Equation 8.3 corresponds to the objective function, which computes the total FI values. Equations 8.4 through 8.6 ensure that these nodes cannot have a well due to the fracture-reorientation constraint angle from the minimum horizontal stress direction, as confirmed in Equation 8.7. Equations 8.7 and 8.9 establish the upper limit of the fractures' half-length. Equation 8.8 ensures that the fractures propagate only in the direction of the minimum horizontal stress. Equations 8.11 through 8.13 establish the continuous range for well origin, connection flow, and fracturing nodes. Equation 8.14 guarantees flow conservation with no sink. Equations 8.15 through 8.18 define three constraints: nonoverlap, length of well, and number of fracture stages.

8.3.4 Stimulated Reservoir Volume Representation

The ideal drainage area of each SRV is shown below in Figure 8.6 as SRV1, SRV2, and SRV3. It is represented in our work through stimulated grids around each created fracture.

Figure 8.7 shows one horizontal well connected with four different SRVs in a form of multistage transverse fractures. Figure 8.8 shows the same number of SRVs connected to four different wells.

8.3.5 Optimization Procedure

The procedure includes four steps, as outlined below:

Step 1: Obtain a detailed 3D map of geomechanical properties of an unconventional shale reservoir using industry-standard sonic techniques.

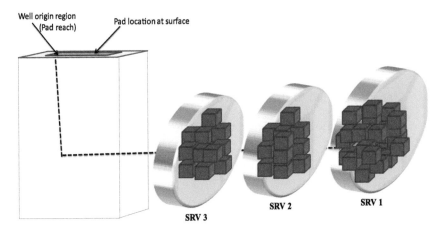

FIGURE 8.6
Drainage volume for different SRVs for different fracture stages branched out from a single horizontal well.

Step 2: Use fracturability index algorithm and previously defined SRV to have 3D distribution of the reservoir quality shale model.

Step 3: Run the field level optimization procedure developed in this work to get the optimum number of wells and fractures, as well as the placement of the wells and fractures.

Step 4: Perform numerical (computational) simulations for the optimum results.

FIGURE 8.7
One horizontal well passes through four different SRVs. Model domain shows the modeled portion of our reservoir along one horizontal well of 10,000 ft. The four different separate regions represent four SRVs of differing volumes.

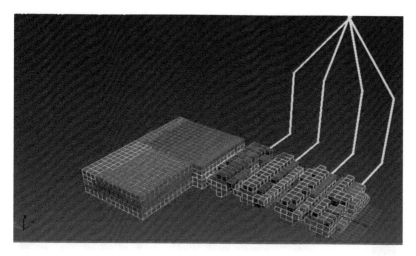

FIGURE 8.8
Combination of four wells and SRVs for the model before optimization, where the blue cells represent the FI below the cutoff.

The sequential steps in this process are shown in the flowcharts provided in Figures 8.3 and 8.4.

The automated process is a direct tool utilizing two software programs: the global optimizer and the reservoir simulator. The first step begins with a spreadsheet input file that is built to facilitate the entry of the FI data or any other input maps. The currently used set of input data maps is generated by use of the correlation published in Alzahabi et al. (2015b).

The number of optimum wells, number of optimum fractures within the wells, and spacing between wells and fractures will be suggested by the use of the approach. It is believed that fracturing the optimum zones will contribute to higher hydrocarbon production from shale and tight formations.

8.4 Model Building

Many wells from Permian Basin are analyzed to help build a representative geochemistry map and mineralogical index. Petrophysical log data from two wells in Wolfcamp were used with geostatistical techniques to construct a detailed geological model that is used here for testing. Multiple relationships of parameters such as porosity, permeability, quartz, clay content, E, and ν were tested in this work to understand Wolfcamp shale. The reservoir model has many 2D layers. The 2D nodes have FI values assigned to them. The commercial reservoir simulators Eclipse and Petrel were used to populate the properties; then the correlation of FI was programmed to generate the quality maps of fracturability index values, denoted by (FI), as shown in Figure 8.9.

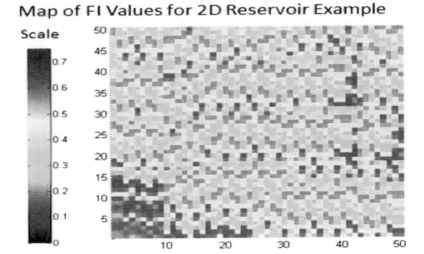

FIGURE 8.9
Fracturability index distribution for the middle layer in the $80 \times 80 \times 5$ model.

The terminology quality map was introduced by Da Cruz et al. (1999) and is commonly used in conventional reservoirs in identifying producing regions. The heterogeneous properties of the reservoir model are represented in the FI values. These input maps serve as input for the optimization developed model.

Quality map generations were applied for each layer in the reservoir model.

The importance of the new model proposed here lies in its simplicity and relative accuracy for the theory described in this work. More importantly, it is based on easily obtainable maps that are increasingly available in contemporary applications of shale characterization. Unlike many of the available SRV prediction tools, the new model does not require information obtained from the real microseismic data. Therefore, the SRV can be estimated before drilling many wells in the reservoir.

Figure 8.10 compares a generated SRV before and after applying the filter of FI = 0.5 on a chosen fracture stage of the shale model.

Figure 8.10 shows the difference in grid-based modeling of one SRV before and after removing the cells that have values of FI < 0.5.

8.4.1 Simulation Model of Well Pad and Stimulated Reservoir Volume Evaluation

To build a fast and accurate mathematically optimized approach to locate wells and SRVs in the shale reservoir model, a model is built with random natural fracture distribution to represent the complexity of shale rock. Figure 8.11 shows a comparison between optimized placements of fracks in terms of SRV versus uniform optimized cases.

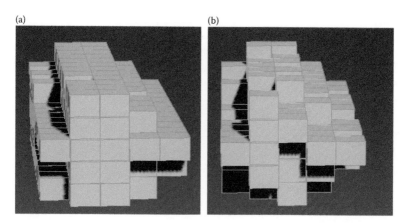

FIGURE 8.10
A comparison between SRV before and after applying the filter of FI; notice that some grids with assigned values of FI < 0.5 were removed in Figure 8.10b. (a) One chosen SRV before applying the filter of FI > 0.5. (b) One SRV after applying the filter of FI > 0.5.

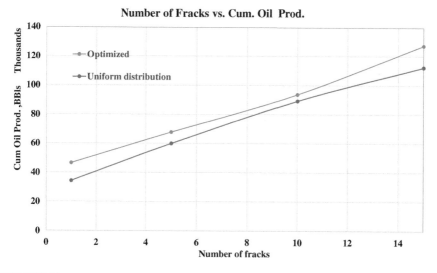

FIGURE 8.11
Optimized placement of the same size SRV versus uniform distribution of SRVs along one horizontal well.

8.5 Results and Discussion

This section is to evaluate the performance of FI via reservoir simulation. The Permian Basin evaluated consists of hundreds of fractures per well. Table 8.1 shows the data ranges used to develop the model.

TABLE 8.1

Properties of the Reservoir Model Used in Validating the Developed Model

Parameters	Minimum Value	Maximum Value
Mineralogical Properties		
Quartz wt. %	6.00	75.0
Calcite wt. %	0.00	84.0
Clay wt. %	3.00	49.0
Pyrite wt. %	0.00	8.00
Petrophysical Properties		
Photoelectric index (P_e), barns/electron	2.61	5.71
Density, ρ_z, g/cc	2.41	2.71
Geomechanical properties		
E, psi	0.38 E6	9.75 E6
ν, ratio	0.02	0.38
Reservoir Properties		
Initial reservoir pressure, psia	5330	
Thickness, ft.	900	
Model dimensions	$80 \times 80 \times 5$	
Cell dimensions, ft.	$30 \times 30 \times 50$	
Porosity, %	Avg.: 9	
Permeability, nanodarcy	Avg.: 208	

As the optimization technique has evolved, statistical algorithms and many software packages have been developed to improve the understanding of fracture stages and the SRV concept, and complex simulations are being implemented to take into account a greater amount of variation in input parameters; however, the importance of considering the power of FI correlation and its predefined cutoffs remains paramount.

The model presented in this work is based on coupling the sweet-spot proxy and optimization tool. It requires input maps of calculated fracturability indices. Contrary to detailed microseismic-based techniques, it requires downhole sensing tools.

8.6 Conclusions and Recommendations

In this chapter, an analysis of placing optimum wells and fractures in the sweet-spot regions of shale reservoirs is presented. These sweet spots are known as grid-assigned high values of FI. The time and number of required stages to reach the objective function was investigated. The fracture and well spacing were assumed to be uneven. SRV size is not equal among stages due to the heterogeneous nature of the rock represented in varied FI values. The advantage of this assumption helps in cost reduction in the placing of wells and fracture stages.

Concluding remarks from this chapter are listed below.

1. In this work, a mathematical optimization approach for the placement of horizontal wells and hydraulic fractures within shale reservoirs was developed. The approach provides a design that gives the optimal predicted stimulation of sweet-spot locations that are identified by the use of the fracturability index. The technique suggests the optimal number of wells and fractures needed in order to drain the shale reservoir by achieving the maximum contact area while respecting the physical and economic constraints.

2. A model that includes the coupling of geomechanical and mathematical optimization was determined for the well data by use of a sophisticated integer programming approach. It is believed that the proposed model arrived at in this analysis is the best of its kind in the industry. A comparison of our proposed model versus published models (although published models are based on other nonoptimal algorithms) shows better results in terms of accuracy in placing fractures. As a final recommendation, more refined models could be proposed in future work involving the collection of more data.

3. Geometric placement of SRVs and hydraulic fracture stages in shale and tight formations should be replaced by coupled approaches of sweet-spot indices and optimization methodologies.

Appendices

Appendix A: Abbreviations

E	Young's modulus
ν	Poisson's ratio
F_{CD}	Dimensionless conductivity
K	Permeability
$\Delta\sigma_h$	Difference between minimum and maximum horizontal stress
FI	Fracturability index
E	Young's modulus, psi
E'	Plane strain modulus, psi
E'_n	Normalized plane strain modulus

ρ	Density, lb./ft^3
$x_{(i,j)}$	X_Y location of a fracture in the reservoir represents the location (i, j) in the shale formation grid
X	Coordinate axis along well path, ft.
Y	Coordinate axis along fracture path, ft.
BHP	Bottomhole pressure, psi
Q_g	Gas flow rate, Mscf/d
c_f	Formation compressibility, psi^{-1}
L_f	Fracture half-length, ft.
P_i	Initial reservoir pressure, psi
L	Well lateral length, ft.
$\Delta X, \Delta Y$	Model grid dimensions in x and y directions, ft.
P_{net}	Net pressure, psi
X_e, Y_e	Rectangular reservoir shape dimensions in x and y directions, ft.
W	Horizontal well
F	Fracture stage
D_{min}	Minimum well spacing, ft.

Appendix B: Definition of the Fracturability Index Used in the Well Placement Process

In this model, every cell is assigned a number based on the calculated FI and geomechanical properties assigned to each cell. The FI consists of a range between 0 and 1. The objective function is achieved through a sum of the total FI values unlocked by the stage from node m to node n and the total FI values unlocked by fracturing at node k.

Appendix C: Geometric Interpretation of Parameters Used in Building the Model

Figure 8C.1 demonstrates five wells attached to one pad. Figure 8C.2 shows one SRV ideal shape and ideal geometry. Figures 8C.3 through 8C.9 list all geometric interpretation of network connections in wells and fractures.

Figure 8C.3 shows a single well designated by W_n originating at node n moving in the direction of the minimum horizontal stress, where X_{mn} shows continuous well connections from node m to n. Figure 8C.4 shows a continuous flow of nodes representing the fracture origin. Figure 8C.5 defines the bounds of each individual valid SRV. Figure 8C.6 differentiates valid versus invalid fracture locations considering minimum fracture distance. Figures 8C.6 and 8C.7 describe the possible geometrical representation of the well path considering other existing SRVs. Figure 8C.8 introduces the main dimensions of one possible SRV, including fracture stage height and width. Figure 8C.9 explains how SRV is represented in our model, whereas discretized SRV is an approximation of SRV ellipse 3D, as represented in Figure 8C.2.

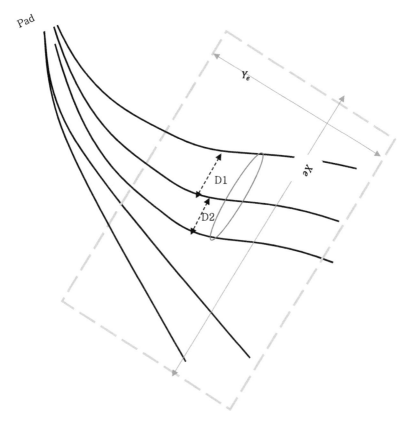

FIGURE 8C.1
Combination of five wells originating from one well pad.

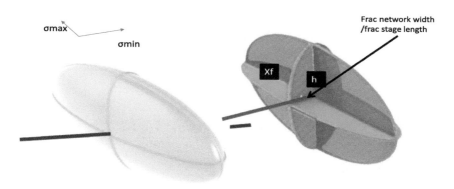

FIGURE 8C.2
Stimulated reservoir volume for one fracture stage.

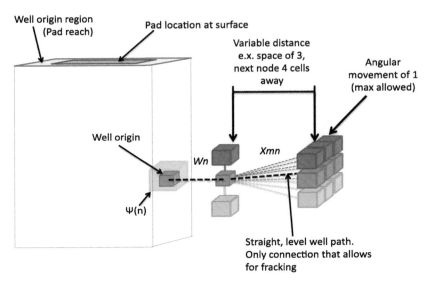

FIGURE 8C.3
Basic network connections in a well.

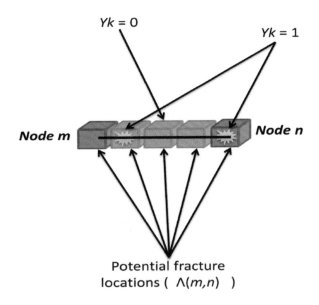

FIGURE 8C.4
Hydraulic fracture stage node representation.

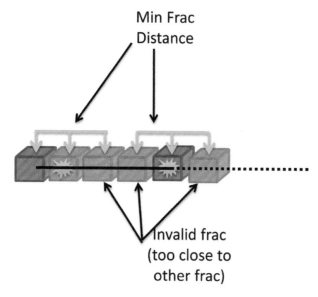

FIGURE 8C.5
Valid versus invalid nodes for fracture stage.

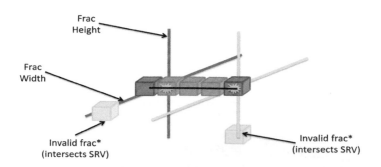

FIGURE 8C.6
Three-dimensional representation of fracture intersection constraints.

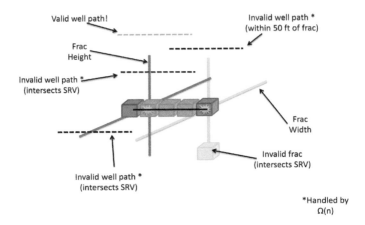

FIGURE 8C.7
Valid paths for neighboring wells.

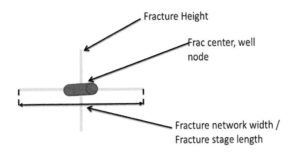

FIGURE 8C.8
Fracture stage parameters.

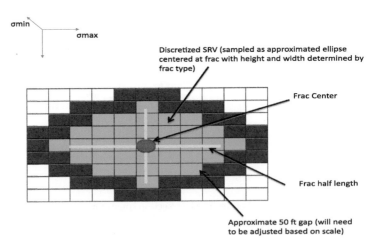

FIGURE 8C.9
Discretized SRV for one fracture stage.

9

Summary, Conclusions, and Recommendations for Future Directions

9.1 Summary

In shale oil and gas resources, multiple parameters control the optimum method of exploitation and development. In the context of recent developments in shale oil and gas optimization, fracturability and brittleness indices (proxies) have attracted greater attention than ever before, with an increased awareness of the need to locate good-quality portions of the rock for placement of horizontal wells and fracture stages. Geochemical properties control the placing of wells, while geomechanical parameters affect the choice of fracture stages. The development of computer speed (CPU) and of numerous algorithms in other industries has made the process significantly easier.

9.2 Conclusions

We studied the relationships among different geomechanical and geochemical and petrophysical properties, and we identified several critical problems in current industry practice, specifically in the design of placing fracture stages and the planning of horizontal wells.

1. The industry trend is to place wells in a uniform manner and fractures equally spaced and distributed along the wellbore.
2. There is no agreement in the industry on a standard brittleness of fracturability scale.
3. There is no real global optimization scale developed for optimizing shale reservoirs.
4. Optimization trials conducted on similar problems were limited to one or two wells and a few simple fractures.

Considering these problems, we introduced a representative fracturability index, an approach for placing fracture stages. FI is a proxy tool that works as an indicator of conditions that favor the creation of fractures in one interval rather than another. Representing both geomechanics and geochemistry, the newly developed index to represent fracturability is a potentially powerful tool for the development of unconventional resources. The index may be utilized for fracture placement and restimulation design. The developed formulation accounts for spacing and cost function when coupled with the proper formulation. A mapped FI would ease the process of the selection of optimum fracture stages through an automated presented approach. A mineralogical index was developed to guide horizontal well placement in shale rock. An integrated fracturability index (IFI) was demonstrated to be an excellent representation of both mineralogical and geomechanical effects. IFI gave an excellent match with minimum horizontal stress, TOC, and differential horizontal stress in placing horizontal wells in the Permian Basin case study. MI, FI, and IFI cutoffs were an excellent addition, especially when coupled with the formulations of the problem.

We studied two different simulation models, conventional and unconventional reservoirs. The data ranges of reservoir characteristics are given in the book.

The proposed coupling of the developed proxies and the mathematically optimized approach for positioning wells and placing fractures leads to efficient well and fracture placement that speeds reservoir depletion. Geological, geochemical, and geomechanical parameters were extensively examined with the aid of literature reviews in the area of operation research and graph theory. In the assessment process, three distinct criteria were designated as follows: excellent rock, good rock, and bad rock. The exact limits of the calculated FI were specified for each of the defined criteria.

The developed method that uses fracturability index maps to optimize fracture placement in new shale fields combines fracturability index mapping with mathematical optimization. The book shows that mathematical formulations and representative proxies should prove an attractive option in placing horizontal wells and designing fracture stages in complex shale plays. The computation time increases for larger fracture half-length and with smaller fracture spacing.

The developed completion optimization algorithm screens data and guides number of fractures, location of fractures and horizontal wells, well spacing, and order (rank) of fracture stages in an optimum time. The computation time of 1368 secs was the optimum in placing two horizontal wells and 80 fracture stages, and achieved the highest objective function. We showed that reservoir permeability, Young's modulus, FI, TOC, and brittleness coefficient affect the fracture locations in a 2D 80×80 model. Therefore, the chosen proxy affects the objective function that governs expected production from the reservoir. The numbers associated with the optimized design rank the possibility of placing fracture stages according to their potential and not fracture creation time.

Furthermore, computational technique performance comparison was investigated for three different standard industry methods (genetic algorithm, simulated annealing, and mixed integer programming) to optimize well placement and solve instances of placing vertical wells in a grid reservoir model. This work considers one formulation of the well placement problem in heterogeneous reservoirs with constraints on interwell spacing. It was found that the existing models and techniques did not adequately predict natural fractures and their propagation under hydraulic pressure. The extensive testing of the three approaches shows IP superiority, particularly in more difficult instances of the tested cases of the conventional reservoirs.

This study focused on the use of linear programming techniques to place horizontal wells and fractures in unconventional resource shale plays. Results demonstrate the superiority of this strategy in reaching the global optimum in many cases tested in this book. Based on the evidence presented herein regarding fracture mechanics–based correlation, we recommend use of the fracturability index in placing fracture stages and use of the mineralogical index for placing horizontal wells. The new model outperforms industry-accepted practice in placing fractures with an increase in total production by 35%.

In addition, the exact limits of good versus bad rock have been identified. A match with different scales used in the industry to reach these limits has been implemented. In lieu of evenly distributed fractures combined with an equally spaced well placement pattern, optimum well locations can be obtained and fracture stage placement achieved in an optimum time through the sets of correlations developed in this book coupled with mathematically introduced methodologies. This method provides a simultaneous determination of the optimal design for horizontal wells and fracture placement. Moreover, in order to obtain results consistent with existing shale plays in industry databases, we established a framework for future shale development that analyzes thickness, depth, TOC, thermal maturity, brittleness, mineral composition, total porosity, net thickness, adsorbed gas, gas content, and geological age. This framework provides for the evaluation of new shale fields and the placement of new shale drilling sites. Statistical similarity and clustering analysis techniques reveal previously unknown relationships among the 12 shales. The proposed approach functions as a metric (Euclidean distance) for quantifying degrees of similarity among the 12 shale plays and identifies operationally approved methods from analogous reservoir development, and the new model outperforms industry-accepted practice as a quick tool in identifying new shales.

9.3 Recommendations for Future Directions

The current model, linked with a fracture simulator, can be modified to include fracture geometry and conductivity. Because it achieved most of the validation through simulation work, the book did not match with a real production logging tool (PLT). Further work is recommended.

The following are recommendations based on this book:

1. A match with production performance per grid of fracture and reservoir would be a necessity.
2. Obtaining a database of production data for the successful shale plays for individual wells and for different gas, condensate, and oil windows must be considered in future work. In addition, the decline in production curve parameters of each major play should be a part of the data.
3. Use of the recommended cutoffs of FI and MI would be helpful in avoiding the transition zone.
4. We recommend extensive use of complex algorithms based on IP in placing different instances of combined wells and fractures for larger data sets than we have in this book with different constraints.
5. Some of these developed correlations are based on Permian Basin shale rock. The mathematically built optimization approach may work with various quality maps for any shale play with consideration of the new constraints.
6. We recommend the use of the integrated fracturability index correlation in placing horizontal wells in shale rock. The use of FI, S1, S2, and DHSR is a promising indication that the new criteria can be applied to future well placement and fracture allocation.
7. We suggest experimentally examining strain rate, stress level, rock mineralogy, Poisson's ratio, and Young's modulus to match the introduced indices. This examination would contribute to understanding of the brittleness and strength of the rock.
8. We suggest extending the developed code to optimize hydraulically induced fracture orientations in a way that respects operational constraints.

Bibliography

AlQahtani, G., Alzahabi, A., Kozyreff, E.I., de Farias, J.R., Soliman, M. 2013. A Comparison between Evolutionary Metaheuristics and Mathematical Optimization to Solve the Wells Placement Problem. *Advances in Chemical Engineering and Science*, 3(4A), 30–36, doi:10.4236/aces.2013.34A1005.

AlQahtani, G.D., Alzahabi, A., Spinner, T., Soliman, M. 2014. A Computational Comparison between Optimization Techniques for Wells Placement Problem: Mathematical Formulations, Genetic Algorithms and Very Fast Simulated Annealing. *Journal of Materials Science and Chemical Engineering*, 2, 59–73. Paper presented at the 4th World Congress on Engineering and Technology (CET 2014), China.

AlQahtani, G., Vadapalli, R., Siddiqui, S., Bhattacharya, S. 2012. Well Optimization Strategies in Conventional Reservoirs. *Proceedings of SPE Saudi Arabia Section Technical Symposium and Exhibition*, Al-Khobar, Saudi Arabia, April 8–11, 2012, 13 p, doi:10.2118/160861-MS

Altindag, R. 2003. Correlation of Specific Energy with Rock Brittleness Concepts on Rock Cutting. *The Journal of the South African Institute of Mining and Metallurgy*, 163–172.

Alzahabi, A., AlQahtani, G., Soliman, M., Bateman, R., Asquith, G., Vadapalli, R. 2015a. Fracturability Index Is a Mineralogical Index, a New Approach for Fracturing Decision. *SPE 2015 Annual Technical Symposium & Exhibition*, Al-Khobar, Saudi Arabia.

Alzahabi, A., Mohamed, A., Soliman, M.Y., Bateman, R., Asquith, G. 2014. Shale Plays Screening Criteria, a Sweet Spot Evaluation Methodology. *Fracturing Impacts and Technologies and Conference*, Lubbock, TX, September 4–5, 2014.

Alzahabi, A., Soliman, M.Y., AlQahtani, G.D., Bateman, R.M., Asquith, G. 2015b. Fracturability Index Maps for Fracture Placement in Shale Plays. *Hydraulic Fracturing Journal*, 2(1), 8–18.

Alzahabi, A., Soliman, M., Asquith, G., Al-Qahtani, G. 2015. Sequencing and Determination of Horizontal Wells and Fractures in Shale Plays: Building a Combined Targeted Treatment Scheme. *Presented at Southwestern Petroleum Short Course Conference*, Lubbock, TX.

Alzahabi, A., Soliman, M., Bateman, R., Asquith, G., Mohamed, A., Stegent, N. 2015. Shale-Gas Plays Screening Criteria: Technology Screens Shale Play Criteria. American Oil and Gas Reporter http://www.aogr.com, http://www.nxtbook. com/nxtbooks/aogr/201504/index.php?startid=79#/74 (Published, April. 2015).

Alzahabi, A., Soliman, M.Y., Thakur, G.C., Trindade, A.A., Stegent, N. 2017. *Horizontal Completion Fracturing Techniques Using Data Analytics: Selection and Prediction*. American Rock Mechanics Association, 51st U.S. Rock Mechanics/ Geomechanics Symposium, 25–28 June, San Francisco, California.

Anderson, D.M., Nobakht, M., Moghadam, S., Mattar, L. 2010. Analysis of Production Data from Fractured Shale Gas Wells. *Presented at the SPE Unconventional Gas Conference*, Pittsburgh, PA. doi:10.2118/131787-MS.

Anderson, T.O., Stahl, E.J. 1967. A Study of Induced Fracturing Using an Instrumental Approach. *Journal of Petroleum Technology*, 19(2), 261–267. doi:10.2118/1537-PA.

API RP 56. 1983. *Recommended Practices for Testing Sand Used in Hydraulic Fracturing Operations*, 1st edition. American Petroleum Institute.

API RP 58 *Recommended Practices for Testing Sand Used in Gravel Packing Operations.*

API RP 60. *Recommended Practices for Testing High-Strength Proppants Used in Hydraulic Fracturing Operations.*

Arab Oil Embargo. 1973. Google free to use images https://www.google.com/search?as_st=y&tbm=isch&as_q=arab+oil+embargo+1973&as_epq=&as_oq=&as_eq=&imgsz=&imgar=&imgc=&imgcolor=&imgtype=&cr=&as_sitesearch=&safe=images&as_filetype=&as_rights

Aronofsky, J.S. 1983. Optimization Methods in Oil and Gas Development. Society of Petroleum Engineers. doi: 10.2118/12295-MS

Azike, O. 2011. Multi-Well Real-Time 3D Structural Modeling and Horizontal Well Placement: An Innovative Workflow for Shale Gas Reservoirs. *Presented at the SPE Eastern Regional Meeting*, Columbus, OH, doi:10.2118/148609-ms

Bai, B., Elgmati, M., Zhang, H. et al. 2013. Rock Characterization of Fayetteville Shale Gas Plays. *Fuel*, 1(05), 645–652.

Bangerth, W., Klie, H., Wheeler, M.F., Stoffa, P.L., Sen, M.K. 2006. On Optimization Algorithms for the Reservoir Oil Well Placement Problem. *Computational Geosciences*, 10(3), 303–319.

Borstmayer, R., Stegent, N., Wagner, A. et al. 2011. Approach Optimizes Completion Design. The American Oil and Gas Reporter, http://www.aogr.com

Bowker, K.A. 2007. Barnett Shale gas production, Fort Worth Basin: Issues and discussion. *AAPG Bulletin*, 91(4), 523–533, doi: 10. 1306/06190606018.

Boyer, C., Kieschnick, J., Rivera, R.S. et al. 2006. Producing Gas from Its Source. *Oilfield Review*, 36–49. https://pdfs.semanticscholar.org/594a/21a3d6f50f0dddaee4269286a8b2df41f61f.pdf

Branagan, P.T., Cipolla, C.L., Lee, S.J., Wilmer, R.H. 1985. Comprehensive Well Testing and Modeling of Pre- and Post-Fracture Well Performance of the MWX Lenticular Tight Gas Sands. *Presented at the SPE/DOE Low Permeability Gas Reservoirs Symposium*, Denver, CO, May 19–22, doi:10.2118/13867-MS

Britt, L.K., Schoeffler, J. 2009. The Geomechanics of a Shale Play: What Makes a Shale Prospective! Paper Presented at the *2009 Regional Meeting*, Charleston, WV.

Bruner, K.R., Smosna, R. 2011. A Comparative Study of the Mississippian Barnett Shale, Fort Worth Basin, and Devonian Marcellus Shale, Appalachian Basin. A report submitted to URSDOE/NETL-2011/1478.

Buller, D., Hughes, S.N., Market, J. 2010. Petrophysical Evaluation for Enhancing Hydraulic Stimulation in Horizontal Shale Gas Wells. *Presented at APE ATCE*, Florence, Italy, September 19–22.

Chaudhary, N.L., Lee, W.J. 2016. Detecting and Removing Outliers in Production Data to Enhance Production Forecasting. *Society of Petroleum Engineers*, doi:10.2118/179958-MS

Cheng, Y., Guo, B., Wei, N. 2015. Prediction of Fracture Population and Stimulated Reservoir Volume in Shale Gas/Oil Reservoirs. *Presented at the SPE Asia Unconventional Resources Conference and Exhibition, Australia*, doi:10.2118/176833-MS

Cheng, Y., McVay, A., Lee, J. 2009. A Practical Approach for Optimization of Infill Well Placement in Tight Gas Reservoirs. *Journal of Natural Gas Science and Engineering*, 1(2009), 165–176, doi:10.1016/j.jngse.2009.10.004

Chorn, L., Stegent, N., Yarus, J. 2014. Optimizing Lateral Lengths in Horizontal Wells for a Heterogeneous Shale Play. *Society of Petroleum Engineers*, doi:10.2118/167692-MS

Cipolla, C.L., Maxwell, S.C., Mack, M.G. 2012. Engineering Guide to the Application of Microseismic Interpretations. *Society of Petroleum Engineers*, doi:10.2118/152165-MS

Cipolla, C.L., Warpinski, N.R., Mayerhofer, M.J., Lolon, E.P., Vincent, M.C. 2008. The Relationship between Fracture Complexity, Reservoir Properties, and Fracture Treatment Design. *Presented at the 2008 SPE Annual Technical Conference*, Denver, CO, September 21–24, doi:10.2118/115769-MS

Clark, J.B., Fast, C.R. 1952. A Multiple-Fracturing Process for Increasing the Productivity of Wells. *American Petroleum Institute*, New York, API-52-104. 104–116.

Cobb, S.L., Farrell, J.J. 1986. Evaluation of Long-Term Proppant Stability. *Presented at the SPE International Meeting on Petroleum Engineering*, Beijing, China, March 17–20, doi:10.2118/14133-MS

Cooke, C.E., Jr. 1973. Conductivity of Fracture Proppants in Multiple Layers. *Journal of Petroleum Technology*, 25(9), 1101–1107.

Cooke, C.E., Jr. 1975. Effect of Fracturing Fluids on Fracture Conductivity. *Journal of Petroleum Technology*, October, 1275–1282.

Cooke, C.E., Jr. 1977. Fracturing With a High-Strength Proppant. *Journal of Petroleum Technology*, 29(10), 1222–1226.

Cullick, A.S., Narayanan, K., Gorell, S. 2005. Optimal Field Development Planning of Well Locations with Reservoir Uncertainty. *Annual Technical Conference and Exhibition*, 1–12, doi:10.2118/96986-ms

Cullick, A.S., Vasantharajan, S., Dobin, M. 2003. Determining Optimal Well Locations from a 3D Reservoir Model. United States Patent US 6,549,879 B1.

Curtis, J.B. 2002. Fractured Shale-Gas Systems. *AAPG Bulletin*, 86(3), 1921–1938.

Da Cruz, P.S., Horne, R.N., Deutsch, C.V. 1999. The Quality Map: A Tool for Reservoir Uncertainty Quantification and Decision Making. *Paper SPE 56578*, Houston, TX.

Dahl, J., Spaid, J., McDaniel, B. et al. 2014. Accessing Shale Asset Success through Applied Reservoir Understanding. *Presented at the Unconventional Resources Technology Conference*, Colorado.

Daneshy, A.A. 2014. Fracture Shadowing: Theory, Applications and Implications. *Society of Petroleum Engineers*, doi:10.2118/170611-MS

Deutsch, C.V. 1998. Fortran Programs for Calculating Connectivity of Three-Dimensional Numerical Models and for Ranking Multiple Realizations. *Computers & Geosciences*, 24(1), 69–76.

Diamond, W.P. 1987. Characterization of Fracture Geometry and Rock Penetrations Associated with Stimulation Treatments in Coalbeds. *Proc., Coalbed Methane Symposium*, Tuscaloosa, AL.

Ding, D.Y. 2008. Optimization of Well Placement Using Evolutionary Algorithms. *SPE EUROPEC/EAGE Annual Conference and Exhibition*.

Downloaded from EIA public website http://2.bp.blogspot.com/-IuvDbOBCXdE/UbTCGhKxzEI/AAAAAAAAElM/paaSqqH3UrI/s640/Black+&+White+Photos+of+Gas+Stations+in+The+1930's+-+1940's+(19).jpg

Dube, H.G., Christiansen, G.E., Frantz, J.H. et al. 2000. The Lewis Shale, San Juan Basin: What We Know Now. *Presented at the Annual Technical Conference and Exhibition*, Dallas, TX, 1–4 October, doi:10.2118/63091-MS

East, L.E., Soliman, M.Y., Augustine, J.R. 2011. Methods for Enhancing Far-Field Complexity in Fracturing Operations. *SPE Production & Operations*, (3), 291–303.

Economides, M.J., Oilgney, R.E., Valkó, P. 2002. *Unified Fracture Design*. Orsa Press, Houston. https://www.amazon.com/Unified-Fracture-Design-Bridging-Practice/dp/0971042705

Ely, J.W., Holditch, S.A., Carter, R.H. 1988. Improved Hydraulic Fracturing Strategy for Fruitland Formation Coal-Bed Methane Recovery San Juan Basin. Guidebook: Geology and Coal Bed Methane Resources of the Northern San Juan Basin, Colorado. and New Mexico, Rocky Mountain Association of Geologists.

Emerick, A.A. et al. 2009. Well Placement Optimization Using a Genetic Algorithm with Nonlinear Constraints. *SPE Reservoir Simulation Symposium*.

Fast, C.R., Holman, G.B., Covlin, R.J. 1977. The Application of Massive Hydraulic Fracturing to the Tight Muddy "J" Formation, Wattenberg Field, Colorado. *Society of Petroleum Engineers*. doi:10.2118/5624-PA, January, 1.

Fisher, M.K., Davidson, B.M., Goodwin, A.K., Fielder, E.O., Buckler, W.S., Steinberger, N.P. 2002. Integrating Fracture Mapping Technologies to Optimize Stimulations in the Barnett Shale. *Presented at the SPE Annual Technical Conference and Exhibition*, San Antonio, TX, September 29–October 2, doi:10.2118/77411-MS

Fisher, M.K., Heinze, J.R., Harris, C.D., Davidson, B.M., Wright, C.A., Dunn, K.P. 2004. Optimizing Horizontal Completions Techniques in the Barnett Shale Using Microseismic Fracture Mapping. *Presented at the SPE Annual Technical Conference and Exhibition*, Houston, TX, September 26–29, doi:10.2118/90051-MS

Fisher, M.K., Wright, C.A., Davidson, B.M. 2002. Integrating Fracturing Mapping Technologies to Optimize Stimulations in the Barnett Shale. *Presented at SPE ATCE*, San Antonio, TX.

Friedrich, M., Monson, G. 2013. Two Practical Methods to Determine Pore Pressure Regimes in the Spraberry and Wolfcamp Formations in the Midland Basin. *Unconventional Resources Technology Conference*.

From Google free to use https://www.bing.com/images/search?view=detailV2&ccid=5%2b%2fNiM0L&id=DCDAFE115DBF081BD8D8D8B84EB2EAD0A39BD4C0&thid=OIP.5-_NiM0LO1IJ-FBY_PisgwHaGu&mediaurl=http%3a%2f%2f2.bp.blogspot.com%2f-IuvDbOBCXdE%2fUbTCGhKxzEI%2fAAAAAAAAElM%2fpaaSqqH3UrI%2fs640%2fBlack%2b%26%2bWhite%2bPhotos%2bof%2bGas%2bStations%2bin%2bThe%2b1930%27s%2b-%2b1940%27s%2b(19).jpg&exph=581&expw=640&q=gas+station+signs+for+the+1940s&simid=608033626104596158&selectedIndex=46&ajaxhist=0

Gray, D., Anderson, P., Logel, J., Delbecq, F., Schmidt, D., Schmid, R. 2012. Estimation of stress and geomechanical properties using 3D seismic data. *First Break*, 30, March 2012, https://www.cgg.com/technicalDocuments/cggv_0000013432.pdf.

Grieser, W.V., Shelley, R.F., Johnson, B.J., Fielder, E.O., Heinze, J.R., Werline, J.R. 2006. Data Analysis of Barnett Shale Completions. *Society of Petroleum Engineers*, doi:10.2118/100674-MS

Griffith, A.A. 1921. The Phenomena of Rupture and Flow in Solids. *Philosophical Trans. Royal Society, London*, A221, 163–198.

Gurobi Team. 2016. http://www.gurobi.com/.

Gutteridge, P.A., Gawith, D.E. 1996. Connected Volume Calibration for Well-Path Ranking. *EPOC'96: European Production Operations Conference*.

Guyagular, B., Horne, R.N. 2001. Uncertainty Assessment of Well Placement Optimization. *SPE Annual Technical Conference and Exhibition*, SPE, doi:10.2118/71625-MS

Hassebroek, W.E., Stegelman, A., Westbrook, S.S.S. 1954. Progress in Sand-oil Fracturing Treatments. *American Petroleum Institute*, New York, API-54-212.

Hassebroek, W.E., Waters, A.B. 1964. Advancements through 15 Years of Fracturing. *Journal of Petroleum Technology*, 760–764.

Henery, F., Trimbitasu, T., Johnson, J. 2011. An Integrated Workflow for Reservoir Sweet Spot Identification. Paper Presented at *2011 Gussow Geoscience Conference*, Banff, Alberta, Canada.

Holditch, S.A. 2006. Tight Gas Sands. *Journal of Petroleum Technology Distinguished Author Series*, 84–90.

Holditch, S.A. 2013. Unconventional Oil and Gas Resource Development—Let's Do It Right. *Journal of Unconventional Oil and Gas Resources*, 1(2), 2–8, doi:10.1016/j.juogr.2013.05.001

Holditch, S.A., Robinsin, B.M., Whitehead, W.S. 1987. The Analysis of Complex Travis Peak Reservoirs in East Texas. *Presented at the SPE/DOE Low Permeability Reservoirs Symposium*, Denver, CO, May 18–19, 1987.

http://www.engineerlive.com/content/23507

ISO 13503 Part 2: Measurements of properties of proppants used in hydraulic fracturing and gravel packing.

Jansen, F.E., Kelkar, M.G. 1996. Exploratory Data Analysis of Production Data. *Society of Petroleum Engineers*, doi:10.2118/35184-MS

Jarvie, D.M., Claxton, B.L. 2002. Barnett Shale Oil and Gas as an Analog for Other Black Shales: Extended Abstract. *AAPG Midcontinent Meeting*, New Mexico.

Jarvie, D.M., Hill, R.J., Rube, T.E., Pollastro, R.M. 2007. Unconventional Shale-Gas Systems: The Mississippian Barnett Shale of North-Central Texas as One Model for Thermogenic Shale-Gas Assessment. *AAPG Bulletin*, 91(4), doi:10.1306/12190606068

Jeu, S.J., Logan, T.L., McBane, R.A. 1988. Exploitation of Deeply Buried Coalbed Methane Using Different Hydraulic Fracturing Techniques in the Piceance Basin, Colorado and San Juan Basin New Mexico. *Presented at the 63rd Annual Technical Conference and Exhibition of the Society of Petroleum Engineers*, Houston, TX, October 2–5, doi:10.2118/18253-MS

Jin, C.J., Sierra, L., Mayerhofer, M. 2013. A Production Optimization Approach to Completion and Fracture Spacing Optimization for Unconventional Shale Oil Exploitation. *Presented at Unconventional Resources Technology Conference*, 12–14 August, Denver, Colorado, USA.

Jin, X., Shah, S.N., Roegiers, J.C. 2014. Fracability Evaluation in Shale Reservoirs— An Integrated Petrophysics and Geomechanics Approach. *Presented at SPE Hydraulic Fracturing Technology Conference*, The Woodlands, TX, February 4–6.

Johnson, R., Wichern, D. 2007. *Applied Multivariate Statistical Analysis*, 6th edition. Pearson Prentice Hall, Upper Saddle River, NJ.

Josse, J., Husson, F. 2016. missMDA: A Package for Handling Missing Values in Multivariate Data Analysis. *Journal of Statistical Software*, 70, 1–31.

Kahraman S., Altindag R. 2004. A Brittleness Index to Estimate Fracture Toughness. *International Journal of Rock Mechanics & Mining Sciences*, 41, 343–348.

Kok, J.C.L., Moon, B., Shim, Y.H. et al. 2010. The Significance of Accurate Well Placement in the Shale Gas Plays. *Presented at the SPE Tight Gas Completion Conference*, San Antonio, TX.

Kowalska, S., Szerszeń, M.M., Cicha-Szot, R. et al. 2013. Correlation of Mineralogical Indices of Brittleness with Acoustic Properties of Rocks in Basins of Different Digenesis. *International Symposium of the Society of Core Analysts*, California.

Le Calvez, J.H., Tanner, K.V., Glenn, S. et al. 2006. Using Induced Microseismicity to Monitor Hydraulic Fracture Treatment: A Tool to Improve Completion Techniques and Reservoir Management. *Presented at the SPE Eastern Regional Meeting*, Canton, OH, October 11–13, doi:10.2118/104570-MS

Lehman, L.V., Jackson, K., Noblett, B. 2016. Big Data Yields Completion Optimization: Using Drilling Data to Optimize Completion Efficiency in a Low Permeability Formation. *Society of Petroleum Engineers*, doi:10.2118/181273-MS

Lewis, R. 2013. *American Association of Petroleum Geologists*. Basic Well Logging School, Course Notes, Golden, CO.

Ma, X., Gildin, E., Plaksina, T. 2015. Efficient Optimization Framework for Integrated Placement of Horizontal Wells and Hydraulic Fracture Stages in Unconventional Gas Reservoirs. *Journal of Unconventional Oil and Gas Resources*, 9, 1–17, doi:10.1016/j.juogr.2014.09.001

MacKay, D. 2003. Chapter 20. An Example Inference Task: Clustering (PDF). In: *Information Theory, Inference and Learning Algorithms*. Cambridge University Press, pp. 284–292.

Manchandda, R., Sharma, M.M., Holzhauser, S. 2014. Time-Dependent Fracture-Interference Effects in Pad Wells. *Presented at Unconventional Resources Conference*, The Woodlands, TX, doi:10.2118/164534-MS.

Maschio, C., Nakajima, L., Schiozer, D. 2008. Production Strategy Optimization Using Algorithm and Quality Map. *Presented at the 2008 SPE Europe/EAGE ATCE*, Italy, doi:10.2118/113483-MS

Mayerhofer, M.J., Lolon, E.P., Warpinski, N.R. et al. 2010. What Is Stimulated Reservoir Volume? *Presented at SPE Shale Gas Production Conference*, Fort Worth, TX, doi:10.2118/119890-MS

Mayerhofer, M.J., Lolon, E., Warpinski, N.R., Cipolla, C.L., Walser, D.W., Rightmire, C.M. 2008. What is Stimulated Rock Volume? *Society of Petroleum Engineers*. January 1. doi:10.2118/119890-MS.

Mayerhofer, M.J., Lolon, E., Warpinski, N.R., Cipolla, C.L., Walser, D.W., Rightmire, C.M. 2010. What Is Stimulated Reservoir Volume? *SPE Production & Operations*, (1), 89–98.

Mayerhofer, M.J., Meehan, D.N. 1998. Waterfracs—Results of 50 Cotton Valley Wells. *Presented at SPE Annual Tech. Conf. & Exhib.*, New Orleans, LA, September 27–30.

Mayerhofer, M.J., Richardson, M.F., Walker, R.N., Meehan, D.N., Oehler, M.W., Browning, R.R. 1997. Proppants? We Don't Need No Proppants. *Presented at the SPE Annual Technical Conference and Exhibition*, San Antonio, TX, October, doi:10.2118/38611-MS

McDaniel, B.W. 1986. Conductivity Testing of Proppants at High Temperature and Stress. *Presented at the California Regional SPE Meeting*, Oakland, CA, April, doi:10.2118/15067-MS

McDaniel, B.W. 1987. Realistic Fracture Conductivities of Proppants as a Function of Reservoir Temperature. *Presented at the SPE/DOE Low Permeability Reservoirs Symposium*, Denver, CO, May 18–19.

McDaniel, B.W. 1990. Hydraulic Fracturing Techniques Used for Stimulation of Coalbed Methane Wells. *SPE Eastern Regional Meeting*, Columbus, OH, October 31–November 2.

McDaniel, B.W. 2007. A Review of Design Considerations for Fracture Stimulation of Highly Deviated Wellbores. *Presented at the SPE Eastern Regional Meeting,* Lexington, KY, October 17–19, doi:10.2118/111211-MS

McDaniel, B.W. 2010. Archives of Wellbore Impression Data from Openhole Vertical Well Fracs in the Late 1960s. *Society of Petroleum Engineers,* doi:10.2118/128071-MS

McDaniel, B.W. 2011. How "Fracture Conductivity is King" and "Waterfracs Work" Can Both Be Valid Statements in the Same Reservoir. *Presented at the Canadian Unconventional Resources Conference,* Calgary, Alberta, Canada, November 15–17, doi:10.2118/148781-MS

McDaniel, B.W., Weaver, J.D. 2012. Fracture Stimulate and Effectively Prop Fracs: The Conductivity Challenges of Liquids Production from Ultralow-Permeability Formations. *Presented at SPE Canadian Unconventional Resources Conference,* Calgary, Canada. October 30–November 1, doi:10.2118/162181-MS

Milestones in the History of U.S. Foreign Relations. "Oil Embargo 1973–1974", https://history.state.gov/milestones/1969-1976/oil-embargo

Miller, C.K., Waters, G.A., Rylander, E.I. 2011. Evaluation of Production Log Data from Horizontal Wells Drilled in Organic Shale. *Presented at the SPE North American Unconventional Gas Conference and Exhibition,* Woodlands, TX.

Modeland, N., Buller, D., Chong, K.K. 2011. Statistical Analysis of Completion Methodology. *Society of Petroleum Engineers,* doi:10.2118/144120-MS

Much, M.G., Penny, G.S. 1987. Long-Term Performance of Proppants Under Simulated Reservoir Conditions. *Society of Petroleum Engineers,* doi:10.2118/16415-MS

Mullen, J., Lowry, J., Nwabuoku, K. 2010. Lessons Learned Developing the Eagle Ford Shale. *Presented at the Tight Gas Completions Conference,* San Antonio, TX, November 2–3, doi:10.2118/138446-MS

Mullen, M., Enderlin, M. 2012. Fracturability Index—More Than Just Calculating Rock Properties. *Presented at the SPE Annual Technical Conference,* San Antonio, TX, January 1, doi:10.2118/159755-MS.

Nakajima, L., Schiozer, D.J. 2003. Horizontal Well Placement Optimization Using Quality Map Definition. Paper 2003–053.

National Petroleum Council. 1967. Impact of New Technology on the U.S. Petroleum Industry 1946–1965. Library of Congress Catalog Card Number: 67–31533. http://www.npc.org/reports/1967-Factors_Affecting_US_Exploration-Development_and_Production-1946-65.pdf

Nemhauser, G., Wolsey, L. 1999. *Integer and Combinatorial Optimization.* 1st edition. Wiley-Interscience.

Nolte, K.G. 1979. Determination of Fracture Parameters From Fracturing Pressure Decline. *Presented at the 34th Annual SPE Tech. Conf. and Exhibition,* Las Vegas, NV. September 23–26, doi:10.2118/8341-MS

Nolte, K.G., Smith, M.B. 1979. Interpretation of Fracturing Pressures. *Presented at the 34th Annual SPE Tech. Conf. and Exhibition,* Las Vegas, NV, September 23–26, doi:10.2118/8297-PA

Nolte, K.G., Smith, M.B. 1981. Interpretation of Fracturing Pressures. *Journal of Petroleum Technology,* 33(9), 1767–1775. doi:10.2118/8297-PA

Northrop, D.A., Sattler, A.R., Mann, R.L., Frohne, K.H. 1984. Current Status of the Multiwell Experiment. *Presented at the SPE/DOE/GRI Unconventional Gas Recovery Symposium,* Pittsburgh, PA, May 13–15.

Oil Field Review Summer. 2013. Stimulation Design for Unconventional Resources. 25(2).

Onwunalu, J.E., Durlofsky, L.J. 2009. Development and Application of a New Well Pattern Optimization Algorithm for Optimizing Large Scale Field Development. *Society of Petroleum Engineers*, doi:10.2118/124364-MS

O'Shea, P.A., Murphy, W.O. 1982. GRI Program for Tight Gas Sands Research. *Presented at the SPE/DOE Unconventional Gas Recovery Symposium of SPE*, Pittsburg, PA, May 16–18.

Padgett, J.C. 1951. Information on Hydrafrac Process. Brill, E.J. (ed.) *Proceedings of the Third World Petroleum Congress*, Section 2. The Hague, Volumes 1-2, 618-622. http://www.npc.org/reports/1968-Impact_of_New_Technology_on_US_Petroleum_Industry-1946-65.pdf

Palmer, L.D., Davids, M.W., Jeu, S.J. 1989. Analysis of Unconventional Behavior Observed during Coalbed Fracturing Treatments. *Proc., Coalbed Methane Symposium*, Tuscaloosa, AL, April, 395–415.

Palmer, I.D., Sparks, D.P. 1989. Measurement of Induced Fractures by Downhole TV Camera in Black Warrior Basin Coal Beds. *Journal of Petroleum Technology*, March 1991, 270–275, 326–328.

Parker, M., Petre, J.E., Dreher, D. 2009. Haynesville Shale: Hydraulic Fracture Stimulation Approach. *Presented at the International Coalbed & Shale Gas Symposium*, Tuscaloosa, AL.

Passey, Q.R., Bohacs, K.M., Esch, W.L., Klimentidis, R., Sinha, S. 2010. From Oil-Prone Source Rock to Gas-Producing Shale Reservoir—Geologic and Petrophysical Characterization of Unconventional Shale-Gas Reservoirs. *Society of Petroleum Engineers*, doi:10.2118/131350-MS

Penny, G.S. 1987. An Evaluation of the Effects of Environmental Conditions and Fracturing Fluids upon the Long-Term Conductivity of Proppants. *Society of Petroleum Engineers*, doi:10.2118/16900-MS

Popa, A.S. 2014. Optimizing the Selection of Lateral Re-Entry Wells through Data-Driven Analytics. *Society of Petroleum Engineers*, doi:10.2118/170702-MS *Proceedings of IAMG, 1997*. Volume 1, pp. 421–426. CIMNE.

Project Gasbuggy; Google free to use images, https://www.google.com/search?as_st=y&tbm=isch&as_q=project+gasbuggy&as_epq=&as_oq=&as_eq=&imgsz=&imgar=&imgc=&imgcolor=&imgtype=&cr=&as_sitesearch=&safe=images&as_filetype=&as_rights=

Project Gasbuggy information downloaded from Wikipedia 6/27/2017.

Pugh, T.D., McDaniel, B.W., Seglem, R.L. 1978. A New Fracturing Technique for Dean Sand. *SPEJ*, 30(2), 167–172, doi:10.2118/6378-PA

R Core Team. 2016. R: A Language and Environment for Statistical Computing, R Foundation for Statistical Computing, Vienna, Austria, https://www.R-project.org

Rafiee, M., Soliman, M.Y., Pirayesh, E. 2012. Hydraulic Fracturing Design and Optimization: A Modification to Zipper Frac. *Presented at the Eastern Regional Meeting*, Kentucky.

Rickman, R., Mullen, M.J., Petre, J. E., Kundert, D. 2008. A Practical Use of Shale Petrophysics for Stimulation Design Optimization: All Shale Plays Are Not Clones of the Barnett Shale. *Presented at SPE Annual Technical Conference and Exhibition (ATCE)*, Denver, CO, September 21–24, doi:10.2118/115258-MS

Roussel, N.P., Sharma, M.M. 2011. Optimizing Fracture Spacing and Sequencing in Horizontal-Well Fracturing. *SPE Production & Operations*, (2), 173–184.

Schamel, S. 2005. Shale Gas Reservoirs of Utah, Survey of an Unexploited Potential Energy Resource. Report Prepared for the Utah Geological Survey, State of Utah, Contract #051845.

Schmid, R., Delbecq, F. 2012. Wide-Azimuth Seismic Enables Better Shale Play Economics, *E and P magazine*.

Schön, J.H. 2011. *Physical Properties of Rocks: A Workbook: Handbook of Petroleum Exploration and Production*. 1st edition. Vol. 8. Elsevier. https://books.google.com/books/about/Physical_Properties_of_Rocks.html?id=L11GJItCu-AC&printsec=frontcover&source=kp_read_button#v=onepage&q&f=false

Sen, M.K., Stoffa, P.L. 1995. *Global Optimization Methods in Geophysical Inversion*. Elsevier.

Shaffner, J.T., Cheng, A., Simms, S., Keyser, E., Yu, M. 2011. The Advantage of Incorporating Microseismic Data into Fracture Models. *Society of Petroleum Engineers*, doi:10.2118/148780-MS

Sharma, R.K., Chopra, S. 2012. An Effective Way to Find Formation Brittleness, AAPG Explorer, online issue 48.

Shebl, M.A., Yalavarthi, R., Nyaaba, C. 2012. The Role of Detailed Petrophysics Reservoir Characterization in Hydraulic Fracture Modeling of Shale AGs Reservoirs. *Presented at the Asia Pacific Oil & Gas Conference and Exhibition*, Perth, Australia, doi:10.2118/160341-MS

Sil, S., Keys, R., Baishali, R., Foster, D. 2013. Methods for seismic fracture parameter estimation and gas filled fracture identification from vertical well log data. U. S. Patent EP2506039A3.

Simon, D.E., McDaniel, B.W., Coon, R.M. 1976. Evaluation of Fluid pH Effects on Low Permeability Sandstones. *Society of Petroleum Engineers*, doi:10.2118/6010-MS

Slocombe, R., Acock, A., Chadwick, C., Wigger, E., Viswanathan, A., Fisher, K., Reischman, R. 2013. Eagle Ford Completion Optimization Strategies Using Horizontal Logging Data. *Unconventional Resources Technology Conference*.

Soliman, M.Y., Dusterhoft, R. 2016. *Fracturing Horizontal Wells*. 1st edition. McGraw-Hill Education, N.p., 5 July 2016. https://www.amazon.com/Fracturing-Horizontal-Wells-Mohamed-Soliman/dp/1259585611

Soliman, M.Y., East, L.E., Adams, D. 2008. Geomechanics Aspects of Multiple Fracturing of Horizontal and Vertical Wells. *SPE Drilling & Completion*, (3), 217–228.

Soliman, M.Y., East, L.E., Augustine, J.R. 2010. Fracturing Design Aimed at Enhancing Fracture Complexity. *Presented at The SPE EUROPEC/EAGE Annual Conference and Exhibition*, Barcelona, Spain, June 14–17.

Soliman, M.Y., Hunt, J.L., Azari, M. 1999. Fracturing Horizontal Wells in Gas Reservoirs. *SPE Prod. & Facilities*, 14(4), SPE 59096.

SPE Monograph Volume 23, Chapter 11

Sarg J.F. 2012. The Bakken—An Unconventional Petroleum and Reservoir System. Technical Report DOE Award No.: DE-NT0005672.

Stegent, N., Wagner, A., Stringer, C. et al. 2012. Engineering Approach to Optimize Development Strategy in the Oil Segment of the Eagle Ford Shale: A Case Study. *Presented at the SPE ATCE*, San Antonio, TX.

Talbi, E.G. 2009. *Metaheuristics: From Design to Implementation*. Vol. 74. John Wiley & Sons.

Tinsley, J.M. 1974. Placing Zones of Solids in a Subterranean Fracture, U. S. Patent No. 3,850,247.

Uhri, D.C., Grand Prairie. *Sequential Hydraulic Fracturing*. U. S. Patent No. 4724905.

U.S. Energy Information Administration—EIA—Independent Statistics and Analysis. (n.d.). Retrieved June 2, 2015, from http://www.eia.gov

van der Baan, M., Calixto, F.J. 2017. Human-Induced Seismicity and Large-Scale Hydrocarbon Production in the USA and Canada. *C3Jour*, 18(7), doi:10.1002/2017GC006915

Vasantharajan, V., Cullick, A.S. 1997. Well Site Selection Using Integer Programming Optimization. *IAMG*, pp. 421–426.

Walker, R.N., Jr., Zinno, R.J., Gibson, J.B., Urbanic, T., Rutledge, J. 1998. Carthage Cotton Valley Fracture Imaging Project—Imaging Methodology and Implications. *SPE Annual Tech. Conf. & Exhib.*, New Orleans, LA, September 27–30.

Walls, J.D., Sinclair, S.W., Devito, J. 2012. Reservoir Characterization in the Eagle Ford Shale Using Digital Rock Methods. *WTGS2012 Fall Symposium*.

Wang, B. 2002. Well site selection algorithm considering geological, economical and engineering constraints. *Thesis*, University of Alberta.

Wang, F.P., Gale, J.F.W. 2009. Screening Criteria for Shale-Gas Systems. *Gulf Coast Association of Geological Societies Transactions*, 59, 779–793.

Wang, G., Carr, T.R. 2013, Prediction and Distribution Analysis of Marcellus Shale Productive Facies in the Appalachian Basin, USA, Presentation at AAPG Annual Convention and Exhibition, Pittsburgh, Pennsylvania, May 19–22.

Warpinski, N.R., Branigan, P.T., Peterson, R.E., Wolhart, S.E. 1999. An Interpretation of M-Site Hydraulic Fracture Diagnostic Results. *Presented at the SPE Rocky Mountain Regional/Low Permeability Reservoirs Symposium and Exhibition*, Denver, CO, April 5–8, doi:10.2118/39950-MS

Warpinski, N.R., Lorenz, J.C., Branagan P.T., Myal, F.R., Gall, B.L. 1991. Examination of a Cored Hydraulic Fracture in a Deep Gas Well. *Presented at the 66th Annual Technical Conference and Exhibit*, Dallas, TX, October 6–9, doi:10.2118/22876-MS

Waters, G.A., Dean, B.K., Downie, R.C. et al. 2009. Simultaneous Hydraulic Fracturing of Adjacent Horizontal Wells in the Woodford Shale. *Presented at the SPE Hydraulic Fracturing Technology Conference*, The Woodlands, TX, January 19–21, doi:10.2118/119635-MS

Wickham, J., Yu, X., McMullen, R. 2013. Geomechanics of Fracture Density. *Presented at the Unconventional Resources Technology Conference*, Denver, CO, August 12–14.

Wilson, K.C., Durlofsky, L.J. 2012. Optimization of Shale Gas Field Devolvement Using Direct Search Techniques and Reduced-Physics Models. *Journal of Petroleum Science and Engineering*, 108(2013), 304–315, doi:10.1016/j.petrol.2013.04.019

Yaalon, D. 1962. Mineral composition of the average shale. *Clays clay Min*. 5, 31–36. http://www.minersoc.org/pages/Archive-CM/Volume_5/5-27-31.pdf

Yeten, B., Durlofsky, L.J., Aziz, K. 2003. Optimization of Nonconventional Well Type, Location, Trajectory. *SPEJ*, 8(3), 200–210, doi:10.2118/86880-PA

Yu, W. et al. 2013. Optimization of Multiple Hydraulically Fractured Horizontal Wells in Unconventional Gas Reservoirs. *Journal of Petroleum Engineering*, 2013, Article ID 151898.

Zemanek, J., Caldwell, R.L., Glenn, E.E., Holcomb, S.V., Norton, L.J., Straus, A.J.D. 1969. The Borehole Televiewer: A New Logging Concept for Fracture Location and Other Types of Borehole Inspection. *Society of Petroleum Engineers*, 21(6), 762–774, doi:10.2118/2402-PA

Zhang, S.C., Lei, X., Zhou, Y.S., Xu, G.Q. 2015. Numerical Simulation of Hydraulic Fracture Propagation in Tight Reservoirs by Volumetric Fracturing. *Petroleum Science*, 12, 674–682, doi:10.1007/s12182-015-0055-4

Zhou, C., Sil, S. August 8, 2013. Fracture identification from azimuthal migrated seismic data. U. S. Patent No. 20130201795.

Index

Printed and bound by CPI Group (UK) Ltd, Croydon, CR0 4YY

01/11/2024

01782623-0003